Mathematical Principles of Optical Fiber Communications

CBMS-NSF REGIONAL CONFERENCE SERIES IN APPLIED MATHEMATICS

JERROLD E. MARSDEN, *Lectures on Geometric Methods in Mathematical Physics*
BRADLEY EFRON, *The Jackknife, the Bootstrap, and Other Resampling Plans*
M. WOODROOFE, *Nonlinear Renewal Theory in Sequential Analysis*
D. H. SATTINGER, *Branching in the Presence of Symmetry*
R. TEMAM, *Navier-Stokes Equations and Nonlinear Functional Analysis*
MIKLÓS CSÖRGO, *Quantile Processes with Statistical Applications*
J. D. BUCKMASTER AND G. S. S. LUDFORD, *Lectures on Mathematical Combustion*
R. E. TARJAN, *Data Structures and Network Algorithms*
PAUL WALTMAN, *Competition Models in Population Biology*
S. R. S. VARADHAN, *Large Deviations and Applications*
KIYOSI ITÔ, *Foundations of Stochastic Differential Equations in Infinite Dimensional Spaces*
ALAN C. NEWELL, *Solitons in Mathematics and Physics*
PRANAB KUMAR SEN, *Theory and Applications of Sequential Nonparametrics*
LÁSZLÓ LOVÁSZ, *An Algorithmic Theory of Numbers, Graphs and Convexity*
E. W. CHENEY, *Multivariate Approximation Theory: Selected Topics*
JOEL SPENCER, *Ten Lectures on the Probabilistic Method*
PAUL C. FIFE, *Dynamics of Internal Layers and Diffusive Interfaces*
CHARLES K. CHUI, *Multivariate Splines*
HERBERT S. WILF, *Combinatorial Algorithms: An Update*
HENRY C. TUCKWELl, *Stochastic Processes in the Neurosciences*
FRANK H. CLARKE, *Methods of Dynamic and Nonsmooth Optimization*
ROBERT B. GARDNER, *The Method of Equivalence and Its Applications*
GRACE WAHBA, *Spline Models for Observational Data*
RICHARD S. VARGA, *Scientific Computation on Mathematical Problems and Conjectures*
INGRID DAUBECHIES, *Ten Lectures on Wavelets*
STEPHEN F. MCCORMICK, *Multilevel Projection Methods for Partial Differential Equations*
HARALD NIEDERREITER, *Random Number Generation and Quasi-Monte Carlo Methods*
JOEL SPENCER, *Ten Lectures on the Probabilistic Method, Second Edition*
CHARLES A. MICCHELLI, *Mathematical Aspects of Geometric Modeling*
ROGER TEMAM, *Navier–Stokes Equations and Nonlinear Functional Analysis, Second Edition*
GLENN SHAFER, *Probabilistic Expert Systems*
PETER J. HUBER, *Robust Statistical Procedures, Second Edition*
J. MICHAEL STEELE, *Probability Theory and Combinatorial Optimization*
WERNER C. RHEINBOLDT, *Methods for Solving Systems of Nonlinear Equations, Second Edition*
J. M. CUSHING, *An Introduction to Structured Population Dynamics*
TAI-PING LIU, *Hyperbolic and Viscous Conservation Laws*
MICHAEL RENARDY, *Mathematical Analysis of Viscoelastic Flows*
GÉRARD CORNUÉJOLS, *Combinatorial Optimization: Packing and Covering*
IRENA LASIECKA, *Mathematical Control Theory of Coupled PDEs*
J. K. SHAW, *Mathematical Principles of Optical Fiber Communications*

J. K. Shaw
Virginia Polytechnic Institute and State University
Blacksburg, Virginia

Mathematical Principles of Optical Fiber Communications

siam
SOCIETY FOR INDUSTRIAL AND APPLIED MATHEMATICS
PHILADELPHIA

10 9 8 7 6 5 4 3 2 1

Library of Congress Cataloging-in-Publication Data

Shaw, J. K.
Mathematical principles of optical fiber communications / J. K. Shaw.
p. cm.-- (CBMS-NSF regional conference series in applied mathematics ; 76)
"Sponsored by Conference Board of the Mathematical Sciences; supported by National Science Foundation."
Includes bibliographical references and index.
ISBN 0-89871-556-3 (pbk.)
1. Optical communications--Mathematics. 2. Fiber optics. I. Conference Board of the Mathematical Sciences. II. National Science Foundation (U.S.). III. Title. IV. Series.

TK5103.592.F52S53 2004
621.382'75'0151 — dc22

2004041717

Contents

Preface

The synergism between the World Wide Web and fiber optics is a familiar story to researchers in digital communications. Fibers are the enablers of the rates of information flow that make the Internet possible. Currently there are transoceanic optical fiber cables transmitting data at rates in the range of 1 terabit per second (1 Tb/s), or 10^{12} bits per second. To put this into perspective, if one imagines that a typical book might occupy 10 megabits, then 1 Tb/s would be equivalent to transmitting 10^5 books per second, or the contents of a respectable university library in a few minutes. No other medium is capable of this rate of transmission at such distances.

With the maturing of mobile portable telephony and the emerging broadband access market, greater fiber transmission capacity will be essential in the early 21st century. Since the demand for more capacity drives the development of new optics-based technologies, fiber optics therefore remains a vibrant area for research. The fact that the basic technology is mature means that the open questions are more sharply focused and permit deeper mathematical content.

What are fibers, and why is fiber transmission superior in high bit-rate, long-distance communications? How is it possible to transmit terabit messages in 1 second across an ocean or continent? Or, for that matter, is this figure actually small relative to some theoretical limit? These are the kinds of questions taken up in this book. As it turns out, the answers are usually in equal parts mathematical and physical. Indeed, the development of fiber systems is one of the most fascinating stories in modern science because it involves the interlinked and parallel advances of a number of scientific disciplines such as lasers, optical detectors, novel manufacturing techniques, and mathematical modeling, including wave propagation theory. Mathematics has been especially critical since the key physical parameters in fiber design are determined by intrinsic mathematical constraints. The purpose of this book is to provide an account of this side of the fiber story, from the basics up to current frontiers of research.

To be sure, there is already a great deal of mathematical research underway involving optical fibers. More often than not, mathematical articles on the subject start from an object such as the nonlinear Schrödinger equation and proceed with analysis from that point, leaving out connections between obtained results and actual physical systems. This may be entirely appropriate, since the omission of physical motivation can simply be a matter of taste. However, the new or aspiring researcher misses an opportunity to see the context in which mathematical questions arise. Providing such a context is one goal of this book. It is written not only for both mathematical and engineering researchers who are new to the fiber optics area, but also for experienced investigators who may add richness to their own backgrounds through a better understanding of context and nomenclature, both physical and mathematical.

To outline the book, the first chapter presents an overview of fundamental concepts, including basics of digital communications, light guidance, fiber construction, and some history. Fiber modes are discussed in Chapter 2, starting with Maxwell's equations, detailing why communications fibers need to be single mode and explaining the consequences. Chapter 3 presents the nonlinear Schrödinger equation, the essential elements of its derivation, and the physical interpretation of its terms. The variational approach to modeling pulse propagation, which has been a critical tool in the understanding of nonlinear fiber optics since the early 1980s, is taken up in the fourth chapter. In Chapter 5 solitons are discussed from the standpoint of inverse scattering. The chapter and book close with some new results on optical soliton formation thresholds.

The chapters are written so as to be as independent as possible. There are occasional references to previous or subsequent material, but these are given mainly to place topics in context. An effort has been made to keep the reading lively, useful, and neither terse nor ponderous. The selected topics are representative, rather than comprehensive. The aims are (1) to provide readers with a means to progress from limited knowledge to the research frontier in the shortest time possible and (2) to offer experienced researchers, who may have a narrower background in either fiber communications engineering or its mathematical infrastructure, with sufficient information, terminology, and perspective to engage in interdisciplinary activity.

This monograph is a result of the NSF-CBMS regional conference on Mathematical Methods in Nonlinear Wave Propagation held at North Carolina A & T State University, May 15–19, 2002. It is a pleasure to thank the conference participants and especially Dominic Clemence, Guoqing Tang, and the other members of the Mathematics Department at A & T for their gracious hospitality. All who may benefit from these pages are also indebted to the Conference Board of the Mathematical Sciences, SIAM, and the National Science Foundation.

Arlington, Virginia
July 2003

Chapter 1

Background and Introduction

This chapter will give an overview of fibers and how they are used in optical communications. The first section covers very basic terminology used in digital lightwave communications, the second discusses some of the physical properties of fibers, and the third gives a historical perspective.

1.1 Optical Signals

As in any communications network, optical systems have transmitters and receivers. Fiber transmitters are typically either semiconductor lasers or light-emitting diodes (LEDs) coupled with modulators so as to produce a train of light pulses which represent digital bits. Receivers usually consist of photodetectors, which operate on the principle that a small electrical current is generated when a photon is absorbed (photoelectric effect), together with processing electronics. The transmitter and receiver are synchronized so as to launch and detect, respectively, the pulses of light in discrete, predetermined timing windows, that is, set intervals in the time variable. The signal itself serves to synchronize transmitter and receiver by feeding the temporal separation of detected pulses into a timing device.

In the binary signal format, which is by far the most common, the transmitter launches a pulse of light into a given window to represent a "1" bit, whereas no pulse is inserted if a "0" bit is desired. Likewise, the photodetector looks into a timing window and records a "1" if a pulse is detected and a "0" otherwise. In practice, there is always some level of signal present, which may be due to optical noise, spreading from other windows, or other effects. Detectors are therefore decision makers, and the 0/1 decision is triggered by a light intensity threshold. In fiber systems the medium between transmitter and receiver is the thin, flexible glass filament, or fiber, that guides the pulses. Lengths of individual fibers can run from centimeters up to hundreds of kilometers. *Pulse propagation* is the term used to describe the process of moving a pulse through the medium from transmitter to receiver.

The format in which every 1 bit corresponds to an individual pulse with vanishing tails completely confined to a timing window is referred to as *return-to-zero* (RZ). In the *non-return-to-zero* (NRZ) binary format, consecutive 1 bits maintain the intensity level without dropping to 0 at the interface between the consecutive windows. A string of N

consecutive 1 NRZ bits surrounded by 0s is thus in reality a pulse N times as wide as a pulse corresponding to an isolated 1. The RZ and NRZ formats as described here are examples of pulse *modulation formats*, of which there are many and varied. Some modulation formats alter the phases on strings of pulses, in which case the receiver obviously must be able to extract phase information.

Pulses are distorted during propagation, by both random and deterministic processes, and the overriding goal of digital communications is to correctly interpret received bits as they were intended. There are many causes for signal degradation, but the principal deterministic ones in fiber are *attenuation, dispersion*, and *nonlinear effects*. Attenuation, or *loss*, is the process of scattering and absorption of energy by the medium itself. In fibers the loss figure is small compared with other systems, which is one of the reasons for the superiority of glass. Attenuation is measured in the logarithmic *decibel*, or dB, scale. A given ratio R_0, such as the ratio of output to input light intensity, is converted to dB scale by the formula

$$R_{dB} = 10\log_{10} R_0.$$

Modern fibers have attenuation figures of about $0.2dB$ per kilometer [Ag1]; that is, $R_{dB} = -0.2dB/Km$. At $1Km$, R_0 is thus about 0.955; that is, less than 5% of initial pulse energy has been lost. To put this into perspective, a dimming of 5% of light in the visible spectrum is undetectable to the human eye; thus, there would be no discernable change in an image viewed through a pure silica window $1Km$ thick. (*Silica* is silicon dioxide, SiO_2, the principle material used in making glass.) Modern transmitters insert pulses at peak power levels of a few tens of milliwatts (mW; $1mW = 10^{-6}W$) per channel, and optical receivers can detect levels lower than 1%, or $-20dB$, of that figure. Under such conditions a pulse could be detected after traveling over $100Km$, with no amplification or boosting of the signal level. In practice, distances between amplifiers vary due to a number of factors. Fiber attenuation is also frequency dependent. The figure of $0.2dB/Km$ is the attenuation minimum [Ag1] for silica and occurs at a wavelength of about 1.55 microns (or micrometers, abbreviated μm, $1\mu m = 10^{-6}m$). Note that frequency ω and wavelength λ are related by $\lambda\omega = 2\pi c$, where $c = 3 \times 10^8 m/s$ is the vacuum speed of light. There are wavelengths where other minima occur, such as dispersion, to be discussed next, but attenuation is the most basic contributor to signal degradation, and for this reason communications fibers operate near $1.55\mu m$, which is in the near infrared of the color spectrum, beyond the visible range. The attenuation versus wavelength curve is rather flat in the vicinity of the minimum; thus a number of different isolated wavelengths, or *channels*, are available for low-loss transmission.

It should be noted that losses also occur from bending and splicing. If a bend in a fiber has a diameter less than a few inches, loss can be significant. There are various methods of splicing fibers together. *Fusion* splicing, for instance, essentially melts ends together and results in losses of as low as $0.01dB$ and negligible backscatter [IEC].

Fiber dispersion arises from the fact that the speed of light in glass is frequency dependent. Optical pulses have frequency components (colors) that are concentrated around a central wavelength, or *carrier*, which, as mentioned, is usually near $1.55\mu m$. However, a continuum of wavelengths is present in any real signal, and the fact that distinct wavelength components travel at different speeds means that pulses pull apart, spread, or disperse in time (unless measures are taken to prevent it, as will be discussed in Chapter 4). In order to understand pulse spreading, it is convenient to have a measure of pulse width, with reference to the time variable. As the word itself indicates, pulses are short bursts of energy. In fibers the energy is in the form of light and of intensity $|f(z,t)|^2$, where $f(z,t)$ is the electromagnetic field envelope at propagated distance z and time t. Here $|f(z,t)|^2$ has the physical dimension of *power* and is the quantity detected in many receivers. The total energy in a pulse is given by the integral $\int_{-\infty}^{\infty} |f(z,t)|^2 dt$. Since $|f(z,t)|^2$ is normally given in mW, then the total energy has units of picosecond milliwatts ($ps - mW$). A picosecond is 10^{-12} seconds. The physical dimensions $ps - mW$ are the same (but the *units* are different) as those that appear on a consumer's electric power bill, namely, kilowatt hours. One $ps - mW$ is also equal to 1 femtojoule ($1fJ = 10^{-15}J$) in mechanical energy units.

Typically, for each distance z, $|f(z,t)|^2$ is localized in the neighborhood of a global maximum and drops quickly to 0 in both time (t) directions (for RZ pulses). The time variable t can be set to *local*, that is, $t = 0$ at the peak or maximum. For such shapes, which are ideally symmetric, it is meaningful to define a *pulse width*. A common metric is the *half-width–half-maximum* (*HWHM*), $T = T(z)$, defined implicitly by $|f(z,T)| = |f(z,0)|/2$ [Ag1]; it is the local time required for the pulse modulus to decrease by half. More generally, T can be defined in terms of a reference level other than 1/2. If $f(z,t)$ is Gaussian in t, it is customary [Ag1] to define T implicitly by the equation $|f(z,T)| = |f(z,0)|/\sqrt{e}$. For example, let $z = 0$ and $f(0,t) = Me^{-\alpha^2 t^2}$ (compare (3.40) and (4.6) below). Then the half-width at $z = 0$ would be $T = 1/(\alpha\sqrt{2})$, that is, $\alpha^2 = 1/(2T^2)$. Then $f(0,t)$ may be written in the more common form $f(0,t) = Me^{-(t^2/2T^2)}$ [Ag1]. Even though this corresponds to the reference level $1/\sqrt{e}$, it is still customary to refer to T as the HWHM.

In current technology, HWHM values for launched pulses are usually in the range of roughly 10 to $100ps$. To illustrate dispersion, there is a rule of thumb to be derived in Chapter 3 (see the more general equation (3.43)) which gives the approximate HWHM T of a pulse as a function of propagation distance z. The rule is

$$T^2 = T_0^2 + \frac{\beta_2^2 z^2}{T_0^2}, \tag{1.1}$$

where T_0 is the initial ($z = 0$) pulse HWHM and β_2 is a constant that measures fiber dispersion to first order. First-generation fibers, in place since the 1980s, operate at wavelengths where β_2 is about $-20ps^2/Km$. Using this value of β_2, and an initial width $T_0 = 25ps$, set $T = 2T_0$ in (1.1) and solve for z to find $z = 31.25Km$. That is, the HWHM pulse width doubles in $31.25Km$ as a result of dispersion. In order to prevent the pulse from spreading and corrupting neighboring timing windows, the initial pulse parameters must therefore be chosen in anticipation of dispersion. Note that it is not sufficient to simply take T_0 to be exceedingly small because T_0 appears in a denominator in (1.1); that is, taking T_0 smaller can have the effect of making the output width *larger*. Dispersion leads to *intersymbol interference* (ISI), which is the interaction between neighboring bits made possible by pulse spreading and encroachment. For example, if in the bit pattern 101 the outside 1s spread and overlap in the 0 bit slot, the tails can combine to form a *ghost pulse* in the middle timing window, which should be void. Pulse spreading also means that the maximum power level becomes smaller as the energy is dispersed across a wider pulse, which complicates the setting of a detection decision threshold.

The *dispersion constant* β_2 is frequency dependent and vanishes at a wavelength of about $1.3\mu m$ [Ag1]. This does not mean that there is no dispersion at $1.3\mu m$, but instead that the dominant contribution is 0. Some early-generation fibers were designed to operate at $1.3\mu m$. The most recently installed fibers are dispersion-shifted fibers [Gof], whose zero dispersion wavelength has been shifted by a design and manufacturing process so that the dispersion minimum is also near the loss minimum at $1.55\mu m$.

The subject of dispersion will be a factor in essentially every topic in this monograph. This preliminary discussion is meant to provide at this early point an overall sense of its role in fiber communications.

Turning to nonlinearities, about all that can be said at this introductory stage, before encountering the nonlinear Schrödinger equation, is that fiber nonlinearities cause the Fourier spectrum of a signal to change. By *spectrum*, or *Fourier spectrum*, one means the support of the Fourier transform of the envelope $f(z, t)$, that is, the set of wavelengths where the transform is not negligibly small. In an ideal, linear fiber the spectrum is invariant during propagation; no new frequencies are created and none are destroyed. When significant nonlinearities are present, the spectrum changes. If new frequencies are generated, then account must be taken of the receiver bandwidth, which is obviously finite, as well as frequency filters in the system. Frequency components that are outside filter or receiver bandwidths are blocked, and consequently portions of the signal are removed. These issues will be discussed in more detail in Chapters 3 and 4.

1.2 Glass Rods

A communications optical fiber, illustrated in Figure 1.1, consists of a glass rod of index of refraction n_2, on the order of 100 or 200 microns in diameter, with a central *core region* of fewer than 10 microns in diameter and slightly raised index of refraction n_1. This is sheathed in a plastic or polymer coating for protection. The outer cylinder of the lower index of refraction n_2 is known as the *cladding region*; the sheathing is also called the *jacket*. Though one thinks of glass as brittle and prone to shatter, these small-diameter rods of pure silica are strong and flexible. Fibers with a discontinuous index change at the core–cladding interface are called *step index fibers*.

Light is confined to the core by *total internal reflection*, which is a consequence of

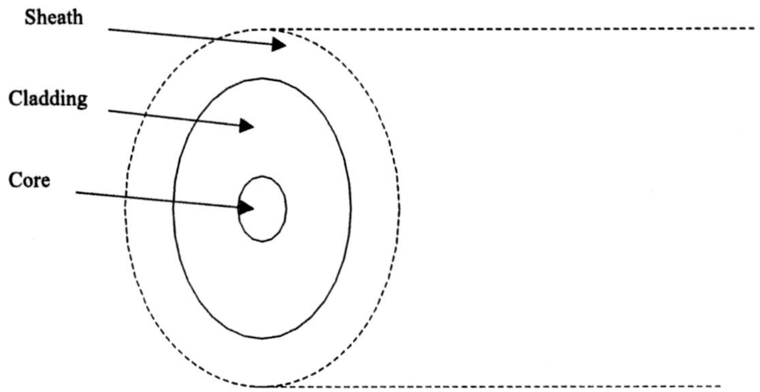

Figure 1.1. *Core, cladding, and jacket of an optical fiber.*

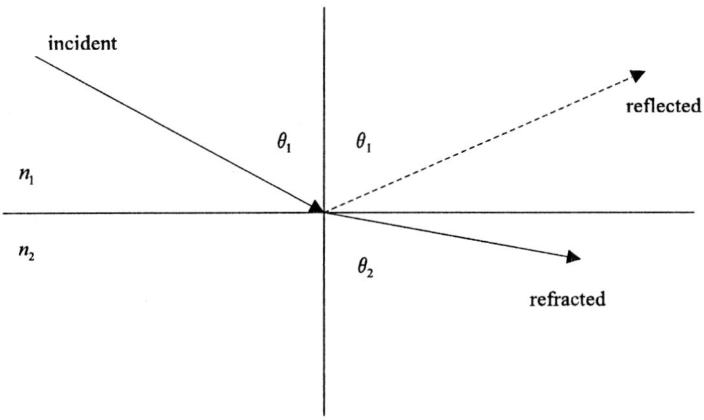

Figure 1.2. *Snell's law.*

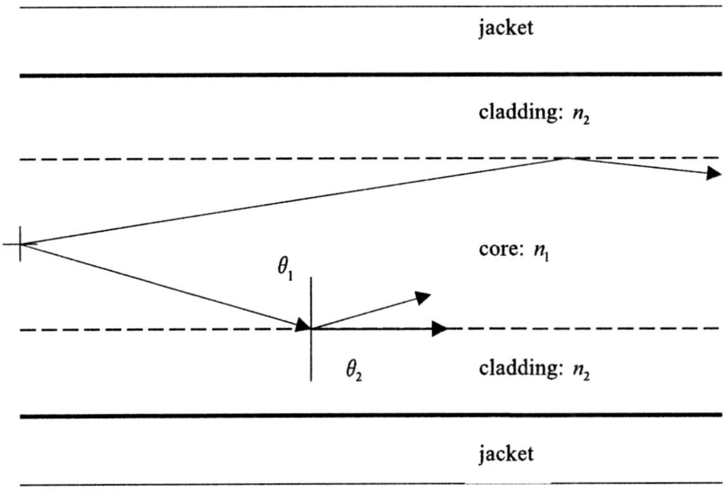

Figure 1.3. *Total internal reflection in a fiber.*

Snell's law. If a light ray strikes the interface between media with indices of refraction n_1 and n_2 at an angle θ_1, as shown in Figure 1.2, then a portion of the light reflects at the incident angle θ_1 and a portion refracts at an angle θ_2 according to $n_1 \sin\theta_1 = n_2 \sin\theta_2$ (Snell's law). In Figure 1.3 (not drawn to scale) a fiber core and cladding are labeled with indices n_1 and n_2, respectively, where $n_1 > n_2$. If in Snell's law θ_1 is sufficiently close to $\pi/2$, then θ_2 is forced to be equal to $\pi/2$ as in Figure 1.3; specifically, this happens when θ_1 attains the *critical angle* $\theta_c = \text{Arcsin}(n_2/n_1)$, at which point all rays are reflected back into the fiber. The figures and discussion here are for planar media, but the same principle applies to rays in a circular core. Since light can be inserted very near the exact center of the core, then rays impinge radially on the core–cladding interface and are reflected or refracted as they would be on striking the tangent line.

The index of refraction of a medium is the ratio of the speed of light in a vacuum to the speed of light in the medium. Fiber cladding usually consists of fused silica, which has an index of about 1.45. The core region is created by doping the silica with germanium [Gof], which raises the index to around 1.47. Using these values for n_1 and n_2 yields a critical angle of $\theta_c = \text{Arcsin}(1.45/1.47) = 80.5^\circ$. The index of refraction actually varies with wavelength; the figure of 1.45 occurs at the standard $1.55\mu m$ carrier.

It is easy to imagine the challenge of fabricating a glass rod with such a small core and with such a small increment in index. Indeed, the core diameter is only a few ten-thousandths of an inch. The cladding outer diameter might run, as mentioned, a few hundred microns or a few thousandths of an inch. The plastic jacket need not be especially thick, since its purpose is to maintain purity rather than add strength; if exposed to the environment and handling, an uncoated silica surface can develop microfractures that lead to weakening. The entire ensemble is thus quite tiny, on the order of a tenth of a millimeter, but surprisingly strong; it is not easy to break a fiber by hand. The thickness of the cladding represents a compromise. It should be thick enough to ensure near-total confinement of light to the core but thin enough to maintain flexibility of the fiber. The jacket may consist of two layers: A soft inner layer is desirable as a cushion and also to ease stripping off of the jacket for splicing and connecting, and the harder outer layer serves as a protector [IEC].

If given a choice, manufacturers would not make fibers with such small core sizes. However, the small size is an intrinsic mathematical constraint arising from the necessity of there being a single eigenvalue for a certain boundary problem. This will be discussed in the next chapter.

The standard way to make fiber is to construct a *preform*, which is a scaled-up version of a fiber, and to draw it into a strand during a molten stage. In the inside vapor deposition (IVD) process, the preform starts as a hollow silica tube. In simple terms, germanium gas is pumped into the inner cylindrical space and the entire tube is heated. In some cases the interior may be placed under a slight vacuum. As in a glass-blowing process, the tube becomes slightly molten and gradually collapses. To prevent bending, the tube is rotated on a lathe during heating. Meanwhile, the inner surface of the tube is coated with a germanium residue that elevates the index of refraction by about 1%. The cooled, collapsed tube is called the *preform*, which is a solid, transparent rod with a small raised-index core of a few microns. Preforms can vary greatly in size. At research laboratories they can be as small as a few centimeters in diameter and a couple of feet in length. Preforms manufactured at the major fiber providers can be sufficiently large and heavy to require mechanical devices for lifting and carrying. An outside vapor deposition (OVD) process deposits both core and cladding sequentially on a "bait" rod [IEC], which is then removed, leaving the preform. If the refractive index changes continuously, the fiber is called a *graded index fiber*.

In the drawing process, which is an impressive science in itself, the preform is fed downward lengthwise into a furnace and pulled into a thin strand with the core remaining at the center. As the fiber is pulled it is coated with the polymer jacket and wrapped onto a spool. The entire drawing, coating, and spooling assembly is called the *draw tower*. The furnace is located at the top, often 20 feet or more above floor level. The strand of fiber runs through a reservoir of liquid polymer for coating, or application of the jacket. There is often a UV radiation chamber located below the polymer applicator to accelerate the curing and drying of the coating. After dropping, drying, and cooling, the fiber is wound onto the rotating spool. Obviously, there are a number of delicate operations running simultaneously and under conditions of precise timing in a draw tower. The speed at which the fiber moves during drawing varies according to the tower type. For smaller laboratory towers the draw speed might be on the order of a meter per second. Industrial draw speeds are often proprietary information but certainly can run in the range of tens of meters per second. The draw speeds are actually controlled by a feedback loop connected to a fiber diameter monitor. That is, the diameter of the drawn fiber can be optically measured and the draw speed automatically adjusted to increase or decrease the diameter.

1.3 Historical Perspective

Communication with electromagnetic waves began in the 1830s with the development of the telegraph; the first trans-Atlantic telegraph cable was installed in 1866. The dot-dash alphabet in Morse code is a form of the binary format. The telephone, invented in 1876, originally operated in analog mode; that is, it used continuously varying electrical current instead of a fixed number of discrete levels (such as 0 and 1). Whether analog or digital, wire pairs still locally connect most homes and businesses to the larger telecommunications network. Once a signal gets out of the home or office and down the street, or from a cellular phone to the base, it can enter the optical network. The copper wire segment of the network is referred to informally as the *last mile*; fiber to the home or office is not widely available because of the expense of optical cable installation. Although the exact form of the next-generation home and office network is not known, the network will eventually be mostly free of wire. Wireless, cellular, and free-space optical communications are prime candidates to replace copper in the 21st century.

From the 1940s through the 1970s, coaxial cable and microwave systems dominated the long-distance industry. However, it had long been understood that optical frequencies have an inherent advantage over cable and microwave since the carrying capacity of any electromagnetic communication channel is proportional to the carrier frequency; that is, the available bandwidth $\Delta\omega$ in a channel is a few percent of the carrier frequency ω [Ag2]. Using the figure of 1% [Ag2], a $10GHz$ microwave carrier can support a channel with HWHM $\Delta\omega = 100MHz$ ($1Hz = 1s^{-1}$ is a standard unit of frequency; $1MHz = 10^6Hz$, $1GHz = 10^9Hz$, $1THz = 10^{12}Hz$). Fourier transforming to the time domain and using $\Delta\omega\Delta t = 1$ ([Ag1]; also see (4.15)) yields a temporal pulse HWHM of $T = 10^{-8}s$, or $10ns$, where ns stands for nanosecond, or $10^{-9}s$. The bit rate $R_b = 1/T$ then becomes $10^8 bits/s$, or $100Mb/s$. A parallel calculation using the $1.55\mu m$ optical carrier with $\omega = 12.15 \times 10^{14}$ yields $\Delta\omega = 12THz$, which is larger than the microwave bandwidth by five orders of magnitude. The reasons coaxial cable, microwave, and satellite systems dominated communications between the 1940s and the 1980s were the lack of low-loss fiber and, until the 1970s, reliable room-temperature lasers. Until 1970, the best grade optical fibers had loss figures in the range of hundreds of dB/Km. But in 1970 the Corning Corporation demonstrated a glass with attenuation below $20dB/Km$ [Hec]. This is equivalent to at least 1% intensity at $1Km$, which is the loss level of unrepeatered copper wire. At that point fiber immediately became competitive. Current fibers have loss figures of $0.2dB/Km$, as mentioned earlier, meaning that signals can propagate over $100Km$ without amplification or regeneration. By comparison, the best coaxial cables require repeater stations spaced no more than a kilometer apart [Ag2].

In 1987 erbium-doped fiber amplifiers became available, which meant that optical signals could be optically amplified; that is, it was no longer necessary to receive, regenerate, and retransmit optical signals. Erbium amplifiers are used because they operate in ranges near $1.55\mu m$. With fiber amplifiers, optical signals can be transmitted practically unlimited distances without electrical intervention. In effect, this means that loss is no longer the dominant impairment for the $1.55\mu m$ carrier; that is, systems can be designed to mitigate other deleterious effects such as dispersion and nonlinearity.

Glass filaments or fibers have a very long history, actually dating back to Roman times [Hec]. However, they were regarded as little more than parlor games and decorations until low-loss fibers became available. The evolution from novelty to communications backbone took years to occur, and in some cases progress had to wait for the development of new technologies such as sources (lasers) and detectors operating at specific wavelengths. In fact, according to some writers [Ag2], the present fibers represent at least the fifth generation. There are several good accounts of the development of optical fibers. The texts [Ag1] and [Ag2] of G. P. Agrawal contain interesting historical discussions, and the recent book of Jeff Hecht [Hec] is a beautiful and detailed history of the subject. References [Sav] and [Den] are historical and general-interest articles on fibers, respectively.

Chapter 2
Fiber Modes

All long-distance communications fibers are *single-mode fibers*, abbreviated *SMF* in the literature. *Single mode* means that the field propagating in the fiber travels at a single *group velocity*, which is the principal part of the wave speed of the pulses. That is, a pulse, which is a superposition of a continuum of frequency components, can be viewed as located and slowly evolving within a moving reference frame, where the speed of the frame is the group velocity. *Multimode* fields are ones in which the field decomposes into components traveling at different group velocities. The dispersion mentioned in Chapter 1 that is caused by different wavelength components traveling at different speeds is called *chromatic* or *intramodal* dispersion and refers to dispersion within a single mode; i.e., intramodal dispersion refers to the changes in pulse shape within the moving time frame. For multimode fields, a pulse can have components corresponding to distinct modes with unequal group velocities, and thus the overall change is much greater; this is called *intermodal* dispersion. Multimode fibers (MMF) are completely undesirable from a long-distance communications standpoint, although there are circumstances, such as in sensing and in local area networks, where MMF can be used (for example, when the propagation distance is only a few kilometers [IEC]). This chapter discusses the way in which fiber geometry leads to the phenomenon of modes, which in turn induces constraints on the physical size of the fiber core.

In order to keep matters as simple as possible, the main presentation will be given for *planar waveguides* and then adapted to circular fibers. All of the principal ideas involved in modes of circular fibers are contained in the simpler theory of planar or *slab* waveguides. The discussion will start at the beginning, namely, Maxwell's equations and boundary conditions at an electromagnetic interface, then proceed to planar waveguides in section 2.2 and circular fibers in section 2.3. The modal patterns obtained in section 2.3 can be greatly simplified by introducing weakly guiding, or linearly polarized, modes. This is covered in section 2.4.

2.1 Maxwell's Equations

Pulses in fibers are physical realizations of electromagnetic (EM) fields that consist of separate electric and magnetic components and are postulated to satisfy Maxwell's equations, namely [Ag1],

$$\begin{aligned}\vec{\nabla} \times \vec{E} &= -\mu_0 \frac{\partial \vec{H}}{\partial t},\\ \vec{\nabla} \times \vec{H} &= \frac{\partial}{\partial t}(\varepsilon_0 \vec{E} + \vec{P}),\\ \vec{\nabla} \bullet \vec{H} &= 0,\\ \vec{\nabla} \bullet \vec{E} &= \vec{\nabla} \bullet \vec{P} = 0,\end{aligned} \tag{2.1}$$

where $\vec{E} = \vec{E}(x, y, z, t)$ and $\vec{H} = \vec{H}(x, y, z, t)$ are the *electric* and *magnetic* fields, respectively. The field $\vec{P} = \vec{P}(x, y, z, t)$ is called the *polarization vector* and will be discussed presently. In (2.1) ε_0 and μ_0 are the free-space *dielectric permittivity* and *magnetic permeability*, respectively, whose specific values [Ish] are unimportant at the moment. One has the relationship [Ag1] $\varepsilon_0\mu_0 = 1/c^2$, c being the vacuum speed of light.

The *polarization vector* $\vec{P}$ in (2.1) corresponds physically to the response of the medium in which the EM field exists. That is, an incident EM field excites the medium, which then responds both linearly and nonlinearly in such a way as to generate a total EM field. In nonlinear fiber optics the polarization vector is assumed to be the sum of linear and nonlinear contributions, namely [Ag1],

$$\vec{P}(x, y, z, t) = \vec{P}_L(x, y, z, t) + \vec{P}_{NL}(x, y, z, t),$$

where specific forms of $\vec{P}_L(x, y, z, t)$ and $\vec{P}_{NL}(x, y, z, t)$ are postulated. The linear part of the response has the form $\vec{P}_L = \varepsilon_1 \vec{E}$, where ε_1 is a constant associated with the propagating medium; in a vacuum, $\varepsilon_1 = 0$.

Chromatic dispersion and the influence of nonlinearities from the term $\vec{P}_{NL}$ turn out to be comparable, that is, they have similar magnitudes and can be compared directly, but both are much smaller effects than intermodal dispersion [Ag1]. For this reason, fiber modes can be defined and characterized without regard to the nonlinear term $\vec{P}_{NL}$, which will consequently be dropped from $\vec{P} = \vec{P}_L + \vec{P}_{NL}$ in (2.1) in the analysis of modes [Ag1]; that is, in (2.1),

$$\varepsilon_0\vec{E} + \vec{P} = \varepsilon_0\vec{E} + \varepsilon_1\vec{E} = \varepsilon\vec{E} \quad \text{and} \quad \vec{P}_{NL} = 0$$

in this chapter.

Here $\varepsilon = \varepsilon_0 + \varepsilon_1$ is the dielectric permittivity of the silica medium. The term $\vec{P}_{NL}$ will become important in the discussion of the nonlinear Schrödinger equation in Chapter 3, where specific forms for the nonlinear response $\vec{P}_{NL}$ and a more general form for the linear term $\vec{P}_L$ are discussed. The form for $\vec{P}_L$ will specialize to $\vec{P} = \varepsilon_1\vec{E}$ as above when the response time of the medium is neglected.

The permittivity ε in a general medium determines the speed of light in the medium by the formula $\varepsilon\mu_0 = 1/v^2$ [Ag1], in analogy to the vacuum relationship $\varepsilon_0\mu_0 = 1/c^2$. The index of refraction, n, of a medium is the ratio of the vacuum speed of light to the speed of light in the medium; that is, $n^2 = c^2/v^2$. From the above expressions, this can be written in the form

$$n^2 = c^2/v^2 = \varepsilon\mu_0/\varepsilon_0\mu_0 = \varepsilon/\varepsilon_0. \tag{2.2}$$

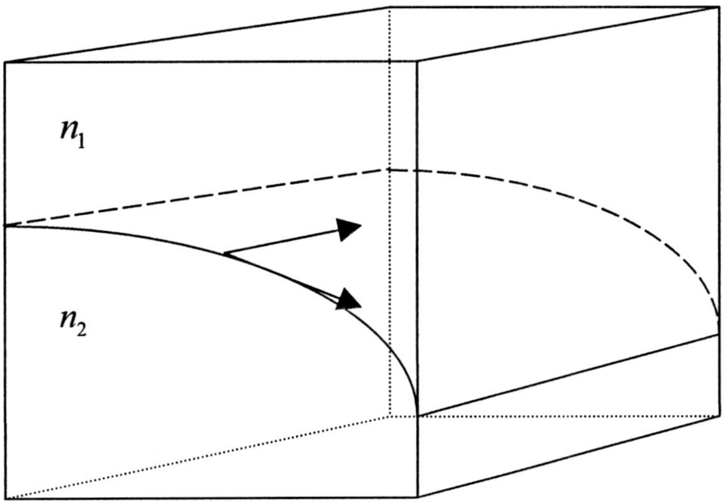

Figure 2.1. *Tangential components at a dielectric interface.*

In view of (2.2) the coefficient ε in (2.1) has a discontinuity at the core–cladding interface, on either side of which the index n takes different values n_1 and n_2 as in Figure 1.3. It turns out in consequence that the optical field cannot be continuous in all components across such an interface. Separation of the field into continuous and discontinuous components is achieved by applying interface conditions at the core boundary. In the case of optical waveguides, the appropriate conditions at an interface are continuity of tangential components [Ish, Mar]. This boundary condition can be justified on the basis of Stokes' theorem and is derived in many books on vector analysis. The requirement is that the components of $\vec{E}$ and $\vec{H}$ that are tangential to the interface be continuous across the interface. Figure 2.1 depicts a generic surface separating a section of waveguide material into regions with indices of refraction n_1 and n_2. The arrows represent components of either $\vec{E}$ or $\vec{H}$ that are tangential to the surface. At any point on the surface the boundary condition requires that the limits of tangential components from either side of the surface be the same.

2.2 Planar Waveguides

Now consider a planar waveguide as illustrated in Figure 2.2. In a slab, planar, or layered waveguide the core is a lateral cross section between $x = -d$ and $x = d$ with index of refraction n_1 sandwiched between regions of index n_2 where $n_1 > n_2$. The physical dimensions are such that the core thickness ($2d$) is many orders of magnitude smaller than either the width, measured along the y axis, or the total depth along the x axis. The variables x and y thus have range $-\infty < x, y < \infty$. The z axis is the direction of pulse propagation. The principle of total internal reflection via Snell's Law confines light to the core region as in Figures 1.2 and 1.3. Fields existing in the planar guide are assumed to be uniform in the y direction, so that all field quantities have vanishing partial derivatives with respect to the y variable ($\partial_y = \partial/\partial y = 0$). In general the refractive indices of the regions on either side

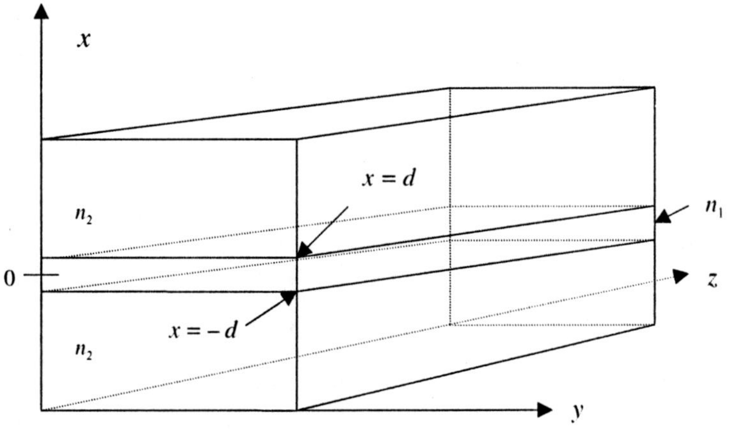

Figure 2.2. *Planar or slab waveguide.*

of the core can differ but are kept equal here for simplicity.

In order to determine modes of (2.1) it will be sufficient to find all solutions of the form

$$\vec{E} = \vec{E}_0(x, y)e^{-i(\omega t - \beta z)}, \qquad \vec{H} = \vec{H}_0(x, y)e^{-i(\omega t - \beta z)}, \tag{2.3}$$

where ω, the central frequency of the signal, is arbitrary and β is to be determined as an eigenvalue parameter. That is, solutions of (2.1) in the form $\vec{E} = \vec{e}(x, y, z)e^{-i\omega t}$ and $\vec{H} = \vec{h}(x, y, z)e^{-i\omega t}$ can be realized as superpositions of solutions of the form (2.3) [Mar]. Values of β will be determined by applying boundary and interface conditions to solutions of (2.1) of the form (2.3).

Introducing the notation $\vec{E}_0(x, y) = (E_x, E_y, E_z)$, $\vec{H}_0(x, y) = (H_x, H_y, H_z)$, where the subscripts designate x, y, and z components, and substituting (2.3) into the second equation of (2.1) gives, by definition of the curl vector,

$$\vec{\nabla} \times \vec{H} = \begin{vmatrix} \vec{i} & \vec{j} & \vec{k} \\ \partial/\partial x & \partial/\partial y & \partial/\partial z \\ H_x e^{-i(\omega t - \beta z)} & H_y e^{-i(\omega t - \beta z)} & H_z e^{-i(\omega t - \beta z)} \end{vmatrix} = \varepsilon \frac{\partial \vec{E}}{\partial t}$$
$$= \varepsilon(-i\omega)\vec{E}_0 e^{-i(\omega t - \beta z)} \tag{2.4}$$

and a similar equation for the curl of $\vec{E}$. Equating components in (2.4) and canceling the common exponential yields the three equations

$$\frac{\partial H_z}{\partial y} - i\beta H_y = -i\omega\varepsilon E_x, \quad -\frac{\partial H_z}{\partial x} + i\beta H_x = -i\omega\varepsilon E_y, \quad \frac{\partial H_y}{\partial x} - \frac{\partial H_x}{\partial y} = -i\omega\varepsilon E_z. \tag{2.5}$$

Imposing the uniformity condition $\partial/\partial y = 0$, (2.5) reduces to

$$i\beta H_y = i\omega\varepsilon E_x, \qquad \frac{\partial H_y}{\partial x} = -i\omega\varepsilon E_z, \qquad -\frac{\partial H_z}{\partial x} + i\beta H_x = -i\omega\varepsilon E_y.$$

Using the other curl equation in (2.1) similarly leads to

$$i\beta E_y = -i\omega\mu_0 H_x, \quad \frac{\partial E_y}{\partial x} = i\omega\mu_0 H_z, \quad \frac{\partial E_z}{\partial x} - i\beta E_x = -i\omega\mu_0 H_y. \tag{2.6}$$

Note that equations (2.5) and (2.6) can be regrouped into sets of independent equations, namely,

$$\begin{aligned} i\beta H_y &= i\omega\varepsilon E_x, \\ \frac{\partial H_y}{\partial x} &= -i\omega\varepsilon E_z, \\ \frac{\partial E_z}{\partial x} - i\beta E_x &= -i\omega\mu_0 H_y, \end{aligned} \tag{2.7}$$

and

$$\begin{aligned} i\beta E_y &= -i\omega\mu_0 H_x, \\ \frac{\partial E_y}{\partial x} &= i\omega\mu_0 H_z, \\ -\frac{\partial H_z}{\partial x} + i\beta H_x &= -i\omega\varepsilon E_y. \end{aligned} \tag{2.8}$$

The total field can thus be decomposed as

$$(E_x, E_y, E_z, H_x, H_y, H_z) = (E_x, 0, E_z, 0, H_y, 0) + (0, E_y, 0, H_x, 0, H_z), \tag{2.9}$$

where the indicated subfields are self-contained by (2.7) and (2.8), respectively. The subfield $(E_x, 0, E_z, 0, H_y, 0)$ is referred to as a *transverse magnetic* (TM) mode or subfield because the magnetic field $\vec{H}$ has no longitudinal (i.e., z) component. Similarly, the subfield $(0, E_y, 0, H_x, 0, H_z)$ is termed *transverse electric* (TE).

It is worth noting that the second two equations in (2.1) have not been used and, in fact, follow from the first two equations for fields of the form (2.3).

An eigenvalue problem will now be solved for the parameter β in the TE case.

TE Modes: $(0, E_y, 0, H_x, 0, H_z)$

Take the derivative of the second equation in (2.8) and substitute from the third equation for the resulting term $\partial H_z/\partial x$; this gives

$$\begin{aligned} \frac{\partial^2 E_y}{\partial x^2} &= i\omega\mu_0 \frac{\partial H_z}{\partial x} = i\omega\mu_0[i\omega\varepsilon E_y + i\beta H_x] \\ &= i\omega\mu_0\left[i\omega\varepsilon E_y + i\beta\left(-\frac{\beta}{\omega\mu_0}E_y\right)\right] = -\omega^2\mu_0\varepsilon E_y + \beta^2 E_y \end{aligned}$$

and finally

$$\frac{\partial^2 E_y}{\partial x^2} + (\omega^2\mu_0\varepsilon - \beta^2)E_y = 0. \tag{2.10}$$

Referring to Figure 2.2, the tangential components are those in the y and z directions. Thus E_y and H_z are continuous in (2.8). By (2.8) $\partial E_y/\partial x$ is also continuous, so that the solution to (2.10) is smooth across the jump discontinuity in ε at the interfaces $x = \pm d$. Note that H_x is also continuous by (2.8) but that $\partial H_z/\partial x$ cannot be continuous since ε has a jump.

Note that either change of variables $\omega \to -\omega$ or $\beta \to -\beta$ leads to the same equation (2.10). Thus the original t and z harmonic assumption $e^{-i(\omega t-\beta z)}$ could be replaced by $e^{i(\omega t-\beta z)}$ with the same result (compare [Mar]).

Equation (2.10) is conventionally written in terms of the wave number k defined by $k = \omega/c = 2\pi/\lambda$ (recalling $\lambda\omega = 2\pi c$). Using (2.2), and recalling that $\varepsilon_0\mu_0 = 1/c^2$ (below (2.1)), one has $n^2k^2 = (\varepsilon/\varepsilon_0)\omega^2\varepsilon_0\mu_0 = \omega^2\varepsilon\mu_0$. In other words (2.10) is the same as

$$\frac{\partial^2 E_y}{\partial x^2} + (n^2k^2 - \beta^2)E_y = 0. \tag{2.11}$$

Equation (2.11) can now be solved subject to the physically plausible conditions that the field E_y should be decaying towards $x = \pm\infty$ and smooth over the transitions at $x = \pm d$. This requires that (2.10) should be oscillatory within $-d \le x \le d$ and nonoscillatory outside this interval, the difference being determined by the value of the index n in core and cladding. It will be convenient to define variables [Ada]

$$K^2 = n_1^2k^2 - \beta^2, \qquad \gamma^2 = \beta^2 - n_2^2k^2. \tag{2.12}$$

Both quantities in (2.12) must be nonnegative (i.e., $n_2^2k^2 \le \beta^2 \le n_1^2k^2$) in order for (2.11) to be oscillatory in the core $-d \le x \le d$ and decaying in the cladding. With the notation of (2.12) the general solution to (2.11) in the core $-d \le x \le d$ is

$$E_y = A_1\cos(Kx) + A_2\sin(Kx), \tag{2.13}$$

while the solution in the cladding is a multiple of $e^{-\gamma(x-d)}$ for $x > d$, and similarly for $x < -d$. In (2.13) A_1 and A_2 are constants. Since the sine and cosine solutions are independent, (2.13) suggests further dividing the TE mode analysis into even and odd parity cases. Consider first the even case.

2.2.1 Even Parity TE Modes: $(0, E_y, 0, H_x, 0, H_z)$

The solution in $-d \le x \le d$ thus has the form $E_y = A\cos(Kx)$ for some constant A. Since the solution will be even symmetric, it is enough to consider only $x > d$, where continuity forces $E_y = A\cos(Kd)e^{-\gamma(x-d)}$, $x > d$. That is, the tangential component E_y is continuous across $x = d$. The other tangential component is H_z, which by (2.8) is given in $-d \le x \le d$ by

$$H_z = \frac{1}{i\omega\mu_0}\frac{\partial E_y}{\partial x} = -\frac{AK}{i\omega\mu_0}\sin(Kx) \tag{2.14}$$

and in $x > d$ by

$$H_z = \frac{1}{i\omega\mu_0}\frac{\partial E_y}{\partial x} = -\frac{A\gamma}{i\omega\mu_0}\cos(Kd)e^{-\gamma(x-d)}. \tag{2.15}$$

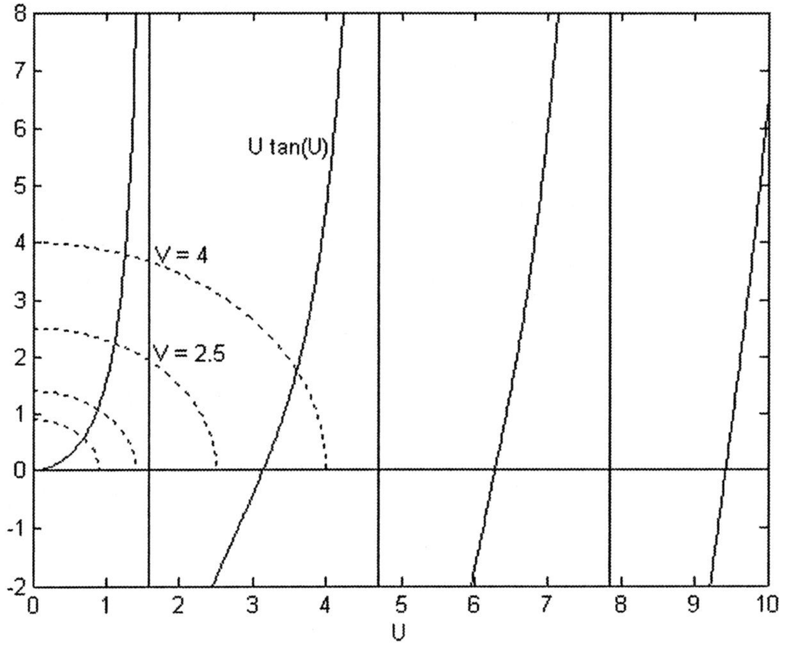

Figure 2.3. *Even TE modes.*

Continuity of the tangential component H_z is guaranteed by equating (2.14) and (2.15) at $x = d$, so as to obtain

$$\tan(Kd) = \frac{\gamma}{K}. \tag{2.16}$$

Noting (2.12) it will be useful to define $V^2 = K^2d^2 + \gamma^2d^2 = k^2d^2(n_1^2 - n_2^2)$ and $U = Kd$ so that $\gamma d = \sqrt{V^2 - U^2}$ together with (2.16) yield the characteristic equation

$$U \tan(U) = \sqrt{V^2 - U^2}. \tag{2.17}$$

The separate sides of (2.17) are graphically illustrated in Figure 2.3 for $V = 0.9$, 1.4, 2.5, and 4.0. The solid curve is the graph of $y = U\tan(U)$ and the dotted curves are plots of $y = \sqrt{V^2 - U^2}$ for $V = 0.9$, 1.4, 2.5, and 4.0. For every $V > 0$ there is at least one intersection, and for $V \geq \pi$ there are at least two intersections. The key observation is that if $V < \pi$ there is exactly one intersection. The root, call it U_0, relates to β through (2.12) as $K^2 = n_1^2k^2 - \beta^2 = U_0^2/d^2$ or $\beta^2 = n_1^2k^2 - (U_0^2/d^2)$. The value of β determines the wave speed in the slab guide for waves of the form (2.3). Note that ω is the frequency and that $k = \omega/c = 2\pi/\lambda$, where λ is the wavelength. By definition of V, the condition $V < \pi$

is equivalent to $(2\pi d/\lambda)\sqrt{n_1^2 - n_2^2} < \pi$ or

$$d < \frac{\lambda}{2\sqrt{n_1^2 - n_2^2}}. \tag{2.18}$$

Using typical values of $n_1 = 1.45$, $n_2 = 1.44$ [Gof] to illustrate, (2.18) becomes $d < 2.94\lambda$. So if the wavelength λ is on the order of $1.5\mu m$, then the condition $V < \pi$ says that the core of the slab in Figure 2.2 can be no thicker than about 9 microns.

Before going on to discuss other types of modes, a point about the unsuitability of multimode waveguides can now be made. Suppose the electric field (a similar analysis holds for the magnetic field) consists of two modes of the form $\vec{E} = \vec{E}_0(x)e^{-i(\omega t - \beta z)}$ for distinct values of β. Components of the field are then superpositions of the form

$$f_1(x)e^{-i(\omega t - \beta_1 z)} + f_2(x)e^{-i(\omega t - \beta_2 z)}.$$

Indeed, the total field can be expressed in the form of a complete orthogonal eigenfunction expansion involving all the modes [Mar]. But assuming that only the two modes propagate, the previous equation can be written as

$$e^{-i(\omega t - \beta_1 z)}(f_1(x) + f_2(x)e^{-i(\beta_1 - \beta_2)z}).$$

At points z along the guide, where $z = \frac{(2n+1)\pi}{(\beta_1 - \beta_2)}$, the above expression reduces to $e^{-i(\omega t - \beta_1 z)}(f_1(x) - f_2(x))$, which is subject to cancellation between $f_1(x)$ and $f_2(x)$ depending on the signs of the individual terms. This is the well-known phenomenon of *destructive interference*. An optical detector looks for a pulse of light in order to record a "1" state, while the absence of light corresponds to the "0" state. Destructive interference can result in holes in the spot of light corresponding to a pulse, or at least diminished intensity sufficient to cause errors. Such interference cannot take place if only one mode is present, and this is the basic idea behind single-mode fibers.

In this same connection it should be mentioned that some optical devices used for sensing are designed so that two modes propagate. Such sensors rely on modal interference patterns to detect changes in the physical environment of the device. For example, a bimodal fiber can be tested to determine an equilibrium interference pattern between two modes. If the fiber is stretched slightly, the pattern changes, and the change can be used to infer the amount of the displacement from equilibrium. In this way multimode fibers can be used to detect, for example, motion or temperature changes.

The other mode types will now be treated, but the analysis is similar and it will be enough to discuss just the main ideas. First, the odd TE modes will be covered.

2.2.2 Odd Parity TE Modes: $(0, E_y, 0, H_x, 0, H_z)$

The solution in $-d \le x \le d$ has the form $E_y = A\sin(Kx)$ for some constant A. Working through the same analysis as above leads to the characteristic equation

$$-U\cot(U) = \sqrt{V^2 - U^2}, \tag{2.19}$$

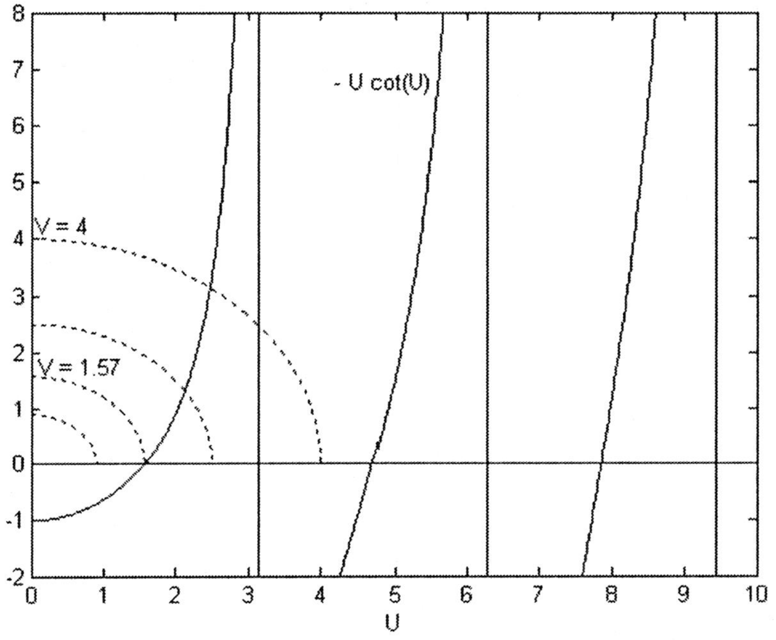

Figure 2.4. *Odd TE modes.*

which is analogous to (2.17). The separate sides of (2.19) are plotted in Figure 2.4 with $V = 0.9$, $\pi/2$, 2.5, and 4.0. The solid curve is the graph of $y = -U\cot(U)$ and the dotted curves are plots of $y = \sqrt{V^2 - U^2}$ for $V = 0.9$, $\pi/2$, 2.5, and 4.0. Note that there is no intersection until $V = \pi/2$ and at least two intersections after $V = 3\pi/2$. Therefore, the lowest order even TE mode propagates alone for $V < \pi/2$.

TM Modes: $(E_x, 0, E_z, 0, H_y, 0)$

Now consider the grouping (2.7). Differentiating the second equation and substituting for $\partial E_z/\partial x$ from the third yields the Sturm–Liouville-type equation

$$\frac{\partial^2 H_y}{\partial x^2} + (n^2k^2 - \beta^2)H_y = 0 \tag{2.20}$$

in complete analogy with (2.11). Define K and γ as in (2.12) and, following (2.13), consider first the even TM modes.

2.2.3 Even TM Modes: $(E_x, 0, E_z, 0, H_y, 0)$

In the region $-d \le x \le d$ the even solution to (2.20) has the form $H_y = A\cos(Kx)$, whereas $H_y = A\cos(Kd)e^{-\gamma(x-d)}$ for $x > d$; it is sufficient to consider only positive values of x by even symmetry. The relevant tangential components are H_y and E_z (see Figures 2.1 and 2.2), where additionally $E_z = -\frac{1}{i\omega\varepsilon}\frac{\partial H_y}{\partial x}$. This case now diverges from (2.14) and (2.15) because the coefficient $1/i\omega\varepsilon$ is discontinuous across $x = \pm d$. In the core region $-d \leq x \leq d$ one has $n = n_1$, with $n = \sqrt{\varepsilon/\varepsilon_0}$. Write $\varepsilon = \varepsilon_1$ in the core and $\varepsilon = \varepsilon_2$ in the cladding. In the core E_z becomes $E_z = \frac{AK}{i\omega\varepsilon_1}\sin(Kx)$ and in the cladding it is $E_z = \frac{-A}{i\omega\varepsilon_2}\cos(Kd)(-\gamma)e^{-\gamma(x-d)}$. Equating these expressions at $x = d$ yields $K\tan(Kd) = \gamma\varepsilon_1/\varepsilon_2$. Recalling $n = \sqrt{\varepsilon/\varepsilon_0}$ leads to the characteristic equation for the even TM modes,

$$U\tan(U) = (n_1^2/n_2^2)\sqrt{V^2 - U^2}, \tag{2.21}$$

where U and V are defined as before.

The odd TM modes similarly have characteristic equation

$$-U\cot(U) = (n_1^2/n_2^2)\sqrt{V^2 - U^2}, \tag{2.22}$$

which is analogous to (2.19). Comparing (2.21) to (2.17) and (2.22) to (2.19) one sees that the only difference between TE and TM modes is in the presence of the term n_1^2/n_2^2. Since $n_1^2 \cong n_2^2$ for most practical communication fibers and planar guides, then the TE and TM modes are degenerate in the sense that they are essentially indistinguishable and inseparable; it would be difficult to excite the lowest order even TE mode without also exciting the lowest order even TM mode. For this reason the TE and TM modes of the same order are grouped together and called *pseudomodes*.

Note that the cutoff V value for the first odd TM mode is slightly larger than $\pi/2$. Thus if $V < \pi/2$, then only the lowest order, or *fundamental*, pseudomode propagates, and it is an even mode. Under this condition the waveguide is termed *single mode*, even though technically there are two physically indistinguishable modes, TE and TM. Using $V < \pi/2$ instead of $V < \pi$ and using the same parameter values as below (2.18), the single-mode condition implies that the slab thickness satisfies $2d < 2.94\lambda$. For wavelengths in the single micron range, the total slab core thickness can therefore be no more than about 4.5 microns. It is very interesting that such a key physical constraint arises from an intrinsically mathematical relationship.

Of course, the lowest order even TE and TM modes have slightly different propagation constants β_1 and β_2, which raises the question of whether the copropagation of these two modes leads to destructive interference, as discussed above in connection with components of the form $e^{-i(\omega t-\beta_1 z)}(f_1(x) - f_2(x))$. However, the interference cycle length $z = (2n + 1)\pi/(\beta_1 - \beta_2)$ is extremely large since $\beta_1 \cong \beta_2$, and this means that the effect is not significant, at least not by comparison with other impairments. In this connection, it should be mentioned that transmitters can be designed so that nearly all the energy is contained in the fundamental mode. Launched pulses are not perfect, and so there will be some contribution from higher order modes, but this contribution is negligible.

To summarize the section, an optical signal in a planar waveguide is basically the field of the fundamental pseudomode, which is an amalgamation of the nearly degenerate, lowest order even TE and TM modes. For communication purposes it is essential that only the fundamental mode propagate, as otherwise the signal can be corrupted by interference between modes. The condition that the waveguide be single mode imposes an a priori constraint on the physical size of the core.

2.3 Circular Fibers

Some key equations will first be derived in the context of general waveguides and then specialized to circular fibers. In this treatment, some technical fine points will be suppressed in order to avoid getting bogged down in details. For example, the final steps leading to (2.34) below will be omitted, as well as the calculation and classification of its roots. References to these derivations will of course be provided.

It will again be sufficient to consider harmonic fields of the form (2.3), since the general field is expandable in an orthogonal series of such terms [Mar] for either planar or circular waveguides.

The separated equations (2.5) and their counterpart continue to hold but cannot be simplified further as was done for the planar case because $\partial/\partial y = 0$ is no longer assumed. That is, a complete classification of modes into TE and TM types does not hold in general; TE and TM modes do exist for circular fibers, but there are other types as well, as will be shown presently. Recall that introducing $\vec{E}_0(x, y) = (E_x, E_y, E_z)$, $\vec{H}_0(x, y) = (H_x, H_y, H_z)$ and using the second curl equation in (2.1) led to (2.5). Likewise, the first curl equation in (2.1) provides an analogous set of equations involving the partial derivatives of (E_x, E_y, E_z). All the resulting equations written together are [Mar]

$$\begin{aligned} &\frac{\partial H_z}{\partial y} - i\beta H_y = -i\omega\varepsilon E_x, \quad -\frac{\partial H_z}{\partial x} + i\beta H_x = -i\omega\varepsilon E_y, \quad \frac{\partial H_y}{\partial x} - \frac{\partial H_x}{\partial y} = -i\omega\varepsilon E_z, \\ &\frac{\partial E_z}{\partial y} - i\beta E_y = i\omega\mu_0 H_x, \quad \frac{\partial E_z}{\partial x} - i\beta E_x = -i\omega\mu_0 H_y, \quad \frac{\partial E_y}{\partial x} - \frac{\partial E_x}{\partial y} = i\omega\mu_0 H_z, \end{aligned} \tag{2.23}$$

which are valid for fields of the form (2.3) in an arbitrary linear waveguide (recalling the assumption $\vec{P}_{NL} = 0$ in this chapter). Now take the first of (2.23), solve for $i\omega\varepsilon E_x$, and replace the term H_y by its value in the fifth equation in (2.23), giving

$$-i\omega\varepsilon E_x = \frac{\partial H_z}{\partial y} + i\beta\left(\frac{-i\beta E_x + \frac{\partial E_z}{\partial x}}{i\omega\mu_0}\right).$$

Letting $K^2 = n^2k^2 - \beta^2$ as in (2.12), but keeping the index n (recall $n^2k^2 = \omega^2\varepsilon\mu_0$) general instead of assuming one of the core or cladding values n_1 or n_2, the previous equation simplifies to

$$E_x = \frac{i}{K^2}\left(\beta\frac{\partial E_z}{\partial x} + \omega\mu_0\frac{\partial H_z}{\partial y}\right).$$

Similarly, the components E_y, H_x, H_y can be solved in terms of the longitudinal components E_z and H_z. Record the four equations as

$$\begin{aligned}
E_x &= \frac{i}{K^2}\left(\beta\frac{\partial E_z}{\partial x} + \omega\mu_0\frac{\partial H_z}{\partial y}\right),\\
E_y &= \frac{i}{K^2}\left(\beta\frac{\partial E_z}{\partial y} - \omega\mu_0\frac{\partial H_z}{\partial x}\right),\\
H_x &= \frac{i}{K^2}\left(\beta\frac{\partial H_z}{\partial x} - \omega\varepsilon\frac{\partial E_z}{\partial y}\right),\\
H_y &= \frac{i}{K^2}\left(\beta\frac{\partial H_z}{\partial y} + \omega\varepsilon\frac{\partial E_z}{\partial x}\right).
\end{aligned} \tag{2.24}$$

In order to solve (2.1) it is therefore sufficient to find the longitudinal components E_z and H_z. Moreover, there are simplified equations for these components. For E_z take the third equation in (2.23) and replace H_x and H_y by the expressions in the third and fourth equations in (2.24). After simplifying, one obtains (the equation for H_z is derived similarly)

$$\frac{\partial^2 E_z}{\partial x^2} + \frac{\partial^2 E_z}{\partial y^2} + K^2 E_z = 0, \qquad \frac{\partial^2 H_z}{\partial x^2} + \frac{\partial^2 H_z}{\partial y^2} + K^2 H_z = 0, \tag{2.25}$$

which are solved subject to interface conditions, as was done with planar waveguides in section 2.2.

Interestingly, all six components of $(\vec{E}, \vec{H})$ satisfy (2.25). To see this, note first that $\vec{\nabla} \bullet \vec{E} = 0$ ($\vec{P}_{NL} = 0$ in the modal analysis of this chapter) from (2.1), and so (2.3) together with $\vec{E}_0(x, y) = (E_x, E_y, E_z)$ implies $\frac{\partial E_x}{\partial x} + \frac{\partial E_y}{\partial y} + i\beta E_z = 0$. Taking the partial with respect to x gives

$$\frac{\partial^2 E_x}{\partial x^2} = -\frac{\partial^2 E_y}{\partial y \partial x} - i\beta\frac{\partial E_z}{\partial x}. \tag{2.26}$$

Now take the partial with respect to y of the last equation in (2.23) to obtain

$$\frac{\partial^2 E_x}{\partial y^2} = \frac{\partial^2 E_y}{\partial x \partial y} - i\omega\mu_0\frac{\partial H_z}{\partial y}. \tag{2.27}$$

Adding (2.26) and (2.27) one obtains $\frac{\partial^2 E_x}{\partial x^2} + \frac{\partial^2 E_x}{\partial y^2} = -i\beta\frac{\partial E_z}{\partial x} - i\omega\mu_0\frac{\partial H_z}{\partial y}$, and the first equation in (2.24) now implies

$$\frac{\partial^2 E_x}{\partial x^2} + \frac{\partial^2 E_x}{\partial y^2} + K^2 E_x = 0,$$

which is the same as the first equation in (2.25), but for E_x instead of E_z. Similarly, the other transverse components satisfy (2.25). Recalling that $K^2 = n^2k^2 - \beta^2$, all components therefore satisfy

$$\frac{\partial^2 E}{\partial x^2} + \frac{\partial^2 E}{\partial y^2} + (n^2k^2 - \beta^2)E = 0, \tag{2.28}$$

where E in (2.28) stands for any of the components $E_x, E_y, E_z, H_x, H_y, H_z$; (2.28) is sometimes called the *scalar wave equation*.

The solution of the first equation of (2.25) will be given next. To keep matters in perspective before going on, the object is to determine values of β so that (2.25) for E_z or H_z, in which $K^2 = n^2k^2 - \beta^2$ with $n = n_1$ in the core and $n = n_2$ in the cladding, has a solution with continuous tangential components at the core–cladding interface and decays in the cladding region.

Since the waveguide is circular it is natural to use polar coordinates $x = r\cos(\varphi)$ and $y = r\sin(\varphi)$, for which (2.25) converts to

$$\frac{\partial^2 E_z}{\partial r^2} + \frac{1}{r}\frac{\partial E_z}{\partial r} + \frac{1}{r^2}\frac{\partial^2 E_z}{\partial \varphi^2} + K^2 E_z = 0. \tag{2.29}$$

Now seek solutions of (2.29) in the separated form $E_z = F(r)G(\varphi)$. It easily follows that $G(\varphi) = e^{i\upsilon\varphi}$ and so

$$E_z = F(r)e^{i\upsilon\varphi}, \tag{2.30}$$

where the physical requirement of 2π periodicity in φ dictates that

$$\upsilon = \pm 1, \pm 2, \ldots; \tag{2.31}$$

that is, υ is an integer. Moreover, the radial part $F(r)$satisfies the Bessel equation

$$\frac{\partial^2 F}{\partial r^2} + \frac{1}{r}\frac{\partial F}{\partial r} + \left(K^2 - \frac{\upsilon^2}{r^2}\right)F(r) = 0. \tag{2.32}$$

Noting Figure 1.1, now assume that the fiber core region, where $n = n_1$, extends over $0 \le r \le d$, where $d > 0$ is the core radius; for $r > d$ the cladding index is $n = n_2$; that is, assume a *step-index* fiber. The bounded solution to (2.32) in the core is of the form $F(r) = AJ_\upsilon(r)$ for some constant A, so that

$$E_z(x, y) = E_z(r, \varphi) = AJ_\upsilon(r)e^{i\upsilon\varphi}. \tag{2.33}$$

In the cladding region the bounded and decaying solution is analogously a multiple of the modified Bessel function K_υ [Ada, Ag1]. At the core–cladding interface the tangential components of $\vec{E}$ are E_z and E_φ (see Figure 2.1), the latter of which can be expressed in terms of E_z using (2.24) after converting the partial derivatives to polar coordinates [Ada]. After equating the respective expressions for E_z and E_φ in the core and cladding, a lengthy calculation yields a characteristic equation in the form [Ada, Ag1]

$$\left(\frac{J_\upsilon'(Kd)}{KJ_\upsilon(Kd)} + \frac{K_\upsilon'(\gamma d)}{\gamma K_\upsilon(\gamma d)}\right)\left(\frac{J_\upsilon'(Kd)}{KJ_\upsilon(Kd)} + \frac{n_2^2}{n_1^2}\frac{K_\upsilon'(\gamma d)}{\gamma K_\upsilon(\gamma d)}\right) = \left(\frac{\upsilon\beta k(n_1^2 - n_2^2)}{dn_1K^2\gamma^2}\right)^2, \tag{2.34}$$

where $k = \omega/c = 2\pi/\lambda$ and K, γ are given by (2.12). Equation (2.34) is solved for β, the propagation constant. That is, (2.34) provides values of β such that solutions of (2.29) of the form (2.33) have continuous tangential components at the interface, are bounded in the entire waveguide, and are decaying in the cladding. Each value of β corresponds to a mode of the fiber.

Note that (2.34) simplifies if $\upsilon = 0$ in (2.31), in which case the field (2.33) is azimuthally uniform. Obviously, the characteristic equation reduces in this case to two subcases, namely,

$$\left(\frac{J'_\upsilon(Kd)}{KJ_\upsilon(Kd)} + \frac{K'_\upsilon(\gamma d)}{\gamma K_\upsilon(\gamma d)}\right) = 0 \tag{2.35}$$

and

$$\left(\frac{J'_\upsilon(Kd)}{KJ_\upsilon(Kd)} + \frac{n_2^2}{n_1^2}\frac{K'_\upsilon(\gamma d)}{\gamma K_\upsilon(\gamma d)}\right) = 0. \tag{2.36}$$

Modes determined by (2.35) have the property that $E_z = 0$ [Ada] and, in analogy with planar guides, are called transverse electric (TE). (It may seem ironic that $E_z = 0$ despite the fact that the eigenvalue problem started by solving for E_z in (2.25); recall, however, that the characteristic equation (2.34) is actually solved for the parameter β without reference to any component.) Modes determined by (2.36) are called transverse magnetic (TM) and satisfy $H_z = 0$. Each equation has infinitely many roots, whose modes are denoted by $\mathrm{TE}_1, \mathrm{TE}_2, \ldots, \mathrm{TM}_1, \mathrm{TM}_2, \ldots$, respectively. Unlike the planar case, where the lowest order TE mode propagates at all wavelengths, the TE_1 mode can be shown to have a *cutoff wavelength* given by the condition [Ada]

$$kd\sqrt{n_1^2 - n_2^2} < c_0, \tag{2.37}$$

where $J_0(c_0) = 0$ in (2.37) and $c_0 \cong 2.405$ is the first zero of the Bessel function J_0. That is, recalling $k = \omega/c = 2\pi/\lambda$, the TE_1 mode does not propagate until the condition $\frac{2\pi d}{\lambda}\sqrt{n_1^2 - n_2^2} \geq c_0$ is satisfied. Since $n_1 \cong n_2$, the same condition serves as the approximate cutoff for the TM modes.

For $\upsilon \neq 0$ there are no solutions of (2.34) corresponding to modes with vanishing longitudinal component, that is, transverse modes. Modes that are neither TE nor TM are called HE and EH modes, depending on a technical condition associated with the roots of (2.34) [Ada]. Corresponding to $\upsilon = 1$ there is a single mode, the HE_{11}, that propagates at all frequencies, where the double subscript notation denotes the first root of (2.34) associated with $\upsilon = 1$. It can be shown [Ada] that all other HE and EH modes are extinct under the condition $\frac{2\pi d}{\lambda}\sqrt{n_1^2 - n_2^2} \geq c_0$. Consequently, (2.37) becomes the criterion that the fiber under consideration is single mode, with only the HE_{11} propagating.

Converting (2.37) to wavelength, where λ is fixed, the single-mode condition is therefore

$$2\pi d\sqrt{n_1^2 - n_2^2} < \lambda c_0.$$

If $n_1 = 1.45$ and $n_2 = 1.44$, then $d < 2.25\lambda$. Taking $\lambda = 1.55$ gives a constraint on the radius d in the form

$$d < 3.5\mu m \tag{2.38}$$

as the single-mode criterion. This is valid for *step index* fibers, which have the simplest refractive index profiles. The single-mode conditions for *graded index* fibers, those with continuously varying index, and for dispersion-shifted fibers, are in general different [IEC].

The phenomenon of destructive interference, discussed in section 2.2 for planar waveguides, also holds for fibers. Communications fibers are therefore single mode, which necessitates (2.37). Condition (2.38), or more generally $2\pi d\sqrt{n_1^2 - n_2^2} < \lambda c_0$, is the primary physical constraint on the core size of optical fibers used in communications.

To summarize the section, light pulses in fibers are realizations of EM fields that propagate in modal patterns. The modes are determined by the fiber geometry. Fibers that carry communication signals are necessarily single mode, as otherwise interference between multiple modes can corrupt the signal. In turn, the single-mode condition leads to a physical constraint on the core size, whose radius can be no more than a few microns.

2.4 Weakly Guiding Modes and Pulses

The characteristic equation (2.34), although providing an exact theoretical account of fiber modes, is unwieldy and not used in practice. Instead, investigators use a scheme devised by Gloge [Glo] which makes an approximation based on the condition $n_1 \cong n_2$ and combines certain nearly degenerate modes so as to form more simply described pseudomodes referred to as *weakly guiding* or *linearly polarized* (LP). Weakly guiding modes will be briefly described in this section, with particular emphasis on the fundamental mode.

Field components corresponding to the modes given by (2.34) are available in explicit form. For all υ, the components in the core region are given, up to a constant multiple, by [Ada]

$$
\begin{aligned}
E_z &= \frac{AKdJ_\nu(Kr)}{kJ_{\nu\pm1}(Kd)}\begin{Bmatrix} \cos(\nu\theta) \\ i\sin(\nu\theta) \end{Bmatrix}, \\
E_x &= \mp Adn_1\frac{J_{\nu\pm1}(Kr)}{J_{\nu\pm1}(Kd)}\begin{Bmatrix} i\cos[(\nu\pm1)\theta] \\ -\sin[(\nu\pm1)\theta] \end{Bmatrix}, \\
E_y &= -Adn_1\frac{J_{\nu\pm1}(Kr)}{J_{\nu\pm1}(Kd)}\begin{Bmatrix} i\sin[(\nu\pm1)\theta] \\ \cos[(\nu\pm1)\theta] \end{Bmatrix},
\end{aligned}
\tag{2.39}
$$

and similar expressions for the H fields, where the upper signs ($\pm$) and upper elements in the brackets correspond to the EH case, and the lower to the HE. Comparing the coefficients of the bracket expressions in (2.39), the significant distinction is K/k in E_z versus n_1 in the other terms. The factor K/k can be written $K/k = \sqrt{n_1^2 - (\beta/k)^2}$, where $-\beta^2/k^2 \leq -n_2^2$ by (2.12), and so $K/k \leq \sqrt{n_1^2 - n_2^2}$. Since $n_1 \cong n_2$, it is the case that $\sqrt{n_1^2 - n_2^2}$ is much less than n_1; recall the values $n_1 = 1.45$, $n_2 = 1.44$ used above (2.38). This is the weakly guiding case, which is defined by the condition $n_1 \cong n_2$ that implies $K/k \cong 0$ in (2.39). Therefore, $E_z \cong 0$ (and similarly $H_z \cong 0$); that is, the modes are all approximately transverse, in that the longitudinal components are negligible. But there is more: for $\upsilon = 1$ in (2.39), $E_x = 0$ in the HE case (lower entry, minus sign). In this approximation the fundamental mode, HE_{11}, thus has only one nonvanishing electric (and magnetic) field component, namely, E_y. The $\vec{E}$ and $\vec{H}$ fields of the fundamental HE_{11} mode are therefore approximately *linearly polarized*;

that is, the longitudinal and one transverse component vanish. In practice, transmitters emit linearly polarized light, which therefore remains approximately linearly polarized because this feature is built into the modal structure.

The conditions $E_z \cong 0$ and $H_z \cong 0$ can be shown to be compatible with (2.24), which would seem otherwise to imply that all components vanish.

Since there is (approximately) only one nonzero component in the weakly guiding case but the characteristic equation (2.34) is derived on the basis of the two tangential components, then the weakly guiding analysis might be expected to extend to the characteristic equation as well. This is indeed the case and it can be shown [Ada] that in the limit $n_1 \rightarrow n_2$, the interface conditions (continuity of tangential components) reduce to continuity of the nonzero scalar component and its directional derivative in the direction normal to the core–cladding interface. This directional derivative is sometimes called the *normal derivative*. The weakly guiding propagation constant β is thus defined as follows. It is a value of β for which (2.28) has a nontrivial solution $E(x, y)$ that is continuous, with a continuous normal derivation, across the core–cladding boundary and decays in the cladding region; compare [SHW]. In this way a complete, though approximate, modal analysis is available that uses only the scalar wave equation (2.28).

The higher order modes associated with (2.34), which will not be of interest in this book, share the feature $E_z \cong 0$ and $H_z \cong 0$. Moreover, modes can be combined to approximately cancel an additional component [Ada]. Specifically, the combinations $\mathrm{HE}_{l+1} \pm \mathrm{EH}_{l-1}$ can be shown to be approximately linearly polarized [Ada]; they comprise the family of *LP modes*. In practice, polarized light launched into a fiber excites one or more LP modes. As mentioned in the previous section, the condition $\frac{2\pi d}{\lambda}\sqrt{n_1^2 - n_2^2} \geq c_0$ guarantees that only the fundamental mode propagates. In the LP designation the fundamental mode HE_{11} is denoted by LP_{01}. The combination of HE_{21}, TE_{01}, and TM_{01} is called the LP_{11} pseudomode; there is a complete family $\{\mathrm{LP}_{jk}\}$ [Ada].

In the analysis of this and the previous section the orientation of transverse coordinate axes is arbitrary, assuming that the fiber is perfectly circular. It is the polarization axis of the inserted light in a fiber that determines the fiber axes. For the fundamental mode, the condition $E_x = 0$ means that the axis of polarization becomes the y axis by definition. It is critical that the modal structure be compatible with the predominant form of light that is actually used in fibers.

Real fibers are not perfectly circular, and if the cross section is not perfectly round, then the propagation constant β will depend on the orientation of axes (which, as mentioned, is determined by the polarization axis of the inserted light). Moreover, the nature of the imperfection in shape changes down the length of the fiber. For example, a fiber may be slightly elliptical and elongated in the vertical direction at the launch point but change to elongation in the horizontal direction further along its length; the cross section undergoes constant minute change. In fact, the principal axis of polarization changes somewhat randomly during transit. Consequently, more than one mode propagates, even if a single mode is launched. With multiple modes propagating, one has modal interference and its attendant dispersion. This type of impairment is called *polarization mode dispersion* (PMD). Although a small effect locally, PMD can become significant over long distances.

As mentioned in Chapter 1, optical signals are pulses of duration in the range 10 to $100ps$. However, the modal structure developed in this chapter is based on the plane wave form (2.3), so these differing views need to be reconciled. At issue is whether the basic modal structure holds for pulses rather than continuous wave fields (2.3). As might be expected, the structure does hold if the pulses as not too short. In fact, if (2.3) is replaced by

$$\vec{E} = \vec{E}_0(x, y) f(t) e^{-i(\omega t - \beta z)}, \qquad \vec{H} = \vec{H}_0(x, y) g(t) e^{-i(\omega t - \beta z)}, \tag{2.40}$$

where $f(t)$ and $g(t)$ are functions of time, then the right side of (2.4) becomes

$$\varepsilon \frac{\partial \vec{E}}{\partial t} = \varepsilon \vec{E}_0 e^{-i(\omega t - \beta z)} (-i\omega f(t) + f'(t)) \tag{2.41}$$

and a similar expression for the curl of the electric field. The left side of (2.4) will likewise contain an additional factor of $f(t)$. If $|\omega|$ is much greater than $f'(t)/f(t)$, then the term containing $f'(t)$ can be neglected in (2.41) and, since the common factor $f(t)$ cancels, then (2.4) does not change even when (2.40) is assumed. Heuristically, the condition $|\omega| \gg f'(t)/f(t)$ holds in practice for several reasons. First, the carrier frequency ω is very large at optical wavelengths. Moreover, pulses are often relatively flat in the region of peak modulus, and thus the derivative term $f'(t)$ is small. Consequently, the modal structure holds at least approximately for the highest intensity portions of a pulse given by (2.40). The extension of (2.3) to (2.40) will be important when the nonlinear Schrödinger equation is considered in the next chapter.

To summarize the section, the fiber mode structure extends to pulses. In the weakly guiding picture, the $\vec{E}$ and $\vec{H}$ fields of the fundamental mode HE_{11} may be considered to have just one nonvanishing component. That is, they are linearly polarized. Combinations of the exact modes arising from (2.34) lead to the family of LP_{jk} modes. The lowest order (fundamental) HE_{11} is also known as the LP_{01} pseudomode.

Chapter 3

Fiber Dispersion and Nonlinearity

The nonlinear Schrödinger equation (NLSE) is the fundamental mathematical device for analyzing nonlinear pulse propagation in fibers. In its simplest form the NLSE includes terms corresponding to two principal physical effects: *dispersion* and *nonlinearity*. The purpose of this chapter is to account for these phenomena. In the first section a derivation of the basic equation will be given. Interpretations of its terms will be covered in the second section, a discussion of the ideal Gaussian pulse model comes in section 3.3, and section 3.4 has some closing remarks.

3.1 The Nonlinear Schrödinger Equation

In its simplest form, the NLSE is [Ag1]

$$i\frac{\partial A}{\partial z} = (\beta_2/2)\frac{\partial^2 A}{\partial t^2} - \gamma|A|^2 A, \tag{3.1}$$

where $A = A(z,t)$ is a function associated with the electric field $\vec{E}$ of an optical pulse in a fiber. Here z is physical length along the longitudinal axis of the fiber and t is time, initialized to the pulse center, while β_2 and γ are constants. The purpose of this section is to derive and begin the analysis of (3.1). Following convention, certain additional terms, which can be important in some situations, have been omitted from (3.1) in the interests of simplicity. The most relevant omitted terms will be discussed presently. As it is, however, (3.1) is the universally accepted basic governing equation for pulse propagation in fibers [Ag1, NeM].

There are higher dimensional analogs of (3.1) [Bou], but propagation in fibers is a phenomenon that occurs naturally in time and one spatial dimension.

The constants β_2 and γ have physical dimensions $(\text{time})^2 \bullet (\text{distance})^{-1}$ and $(\text{distance})^{-1} \bullet (\text{power})^{-1}$, respectively, while $|A(z,t)|^2$ has dimension $(\text{power})^{+1}$. Typically, β_2 is denominated in units of ps^2/Km, γ is in $Km^{-1}mW^{-1}$, and $|A(z,t)|^2$ is in mW; $A(z,t)$ is not assigned units, which would cancel from (3.1) in any event, and so is *dimensionless*.

The initial and boundary conditions associated with (3.1) are that $A(0, t)$ is prescribed and (3.1) is solved subject to $|A(z,t)| \to 0$, $|t| \to \infty$. Physically speaking, pulses are smooth and have compact support, at least when random noise effects are ignored.

The NLSE (3.1) is universal for two reasons: (i) it accurately predicts the behavior of observed optical pulses and (ii) it can be meaningfully derived. The latter is meant in the sense that one can start with Maxwell's equations (2.1), which are postulates, hypothesize a physically reasonable expression for fiber nonlinearity, and deduce (3.1) after performing a set of plausible mathematical steps. That philosophy will be followed here and it seems fitting that some derivation should be given. Other accounts have been provided in [BlW], which is mainly a time domain approach, and in [Fra], which argued in (and in favor of) the frequency domain. An excellent derivation has been provided by Menyuk [Men], who takes into account coupling between differently polarized modes in an analysis based on dominant length scales. The result is a set of coupled equations (the *Manakov* equations) which specialize to (3.1) under conditions when polarization effects, loss, and higher order dispersion can be ignored.

Before discussing the origins of (3.1), note first that it is entirely in the time domain. It will be seen in this chapter that nonlinear effects are more inherently frequency domain phenomena; that is, they are more naturally described in terms of the frequency variable of the Fourier transform. Some insight into this can be provided by solving (3.1) in the two ideal cases where $\gamma = 0$ (pure linear) and $\beta_2 = 0$ (pure nonlinear), respectively. This is done next.

3.1.1 Ideal Linear Case

If $\gamma = 0$, then (3.1) can be solved by Fourier transform. In this case, if $\hat{A}(z,\omega) = \int_{-\infty}^{\infty} A(z,t)e^{i\omega t}dt$ denotes the transform, then (3.1) is equivalent to

$$\partial \hat{A}(z,\omega)/\partial z = i\beta_2\omega^2\hat{A}(z,\omega)/2,$$

whose solution is

$$\hat{A}(z,\omega) = \hat{A}(0,\omega)e^{i\beta_2\omega^2 z/2}. \tag{3.2}$$

Then $|\hat{A}(z,\omega)| = |\hat{A}(0,\omega)|$ for all z; that is, the Fourier transform has constant modulus, and, in particular, magnitudes of frequency components are not changed during propagation.

To comment on nomenclature, ω is called the *frequency variable* or *frequency component* and is equivalent to wavelength λ through $\lambda\omega = 2\pi c$. In the Fourier transform relationship $\hat{f}(\omega) = F[f(t)](\omega) = \int_{-\infty}^{\infty} f(t)e^{i\omega t}dt$, the modulus or amplitude $|\hat{f}(\omega)|$ measures the strength of the frequency component ω. One speaks of *time domain* when discussing $f(t)$ and *frequency domain, Fourier domain*, or *spectral domain* in connection with $\hat{f}(\omega)$. The *width, time domain width*, or sometimes *temporal width* means the pulse width of the function $|f(t)|$. The *spectral width* refers to the width of $|\hat{f}(\omega)|$. The *spectrum* is the set of numerically significant frequency components, that is, values of ω where $|\hat{f}(\omega)|$ is not negligibly small.

It is easy to show that a function $A(z,t)$ whose transform satisfies (3.2) broadens essentially linearly with z in the time domain. Indeed, suppose that $f(z,t)$ is a function that satisfies $\hat{f}(z,\omega) = \hat{f}(0,\omega)e^{ib(\omega)z}$ for some function $b(\omega)$ such as the exponent in (3.2). Let $\sigma(z)$ and $\tilde{\sigma}(z)$ be the *root mean square* (RMS) time and spectral widths for arbitrary z, that is,

$$\sigma^2(z) = \int_{-\infty}^{\infty} \tau^2 |f(z,\tau)| d\tau, \qquad \tilde{\sigma}^2(z) = \int_{-\infty}^{\infty} \omega^2 |\hat{f}(z,\omega)| d\omega, \tag{3.3}$$

where f and $\hat{f}$ denote a Fourier transform pair. The RMS width is a standard measure of the width of a pulse, curve, or distribution [Ag1]. Then

$$\sigma^2(z) = \int_{-\infty}^{\infty} \tau^2 f(z,\tau) f^*(z,\tau) d\tau, \tag{3.4}$$

where the asterisk denotes the complex conjugate, and since the transform of $\tau^2 f(z,\tau)$ is $-\partial_\omega^2 \hat{f}(z,\omega)$, where $\partial_\omega = \partial/\partial\omega$, then Parseval's identity implies

$$\sigma^2(z) = -(1/2\pi) \int_{-\infty}^{\infty} \hat{f}''(z,\omega) f^*(z,\omega) d\omega.$$

From $\hat{f}(z,\omega) = \hat{f}(0,\omega)e^{ib(\omega)z}$ one obtains

$$\begin{aligned} \hat{f}''(z,\omega) = [&\hat{f}''(0,\omega) + ib''(\omega)z\hat{f}(0,\omega) + 2ib'(\omega)\hat{f}'(0,\omega)z \\ &- (b'(\omega))^2 \hat{f}(0,\omega)z^2]e^{ib(\omega)z}, \end{aligned}$$

where the prime denotes differentiation with respect to ω. Inserting this into (3.4) and integrating yields

$$\sigma^2(z) = c_0 + c_1 z + c_2 z^2, \tag{3.5}$$

where the coefficients are integrals involving $\hat{f}$ and its derivatives. The squared RMS width (3.5), defined in (3.3), corresponding to a solution of (3.1) in the pure linear case ($\gamma = 0$), is thus quadratic in z. To summarize, the constant β_2 is responsible for the spreading of a pulse in the time domain in the pure linear case, while the spectral width is invariant.

3.1.2 Pure Nonlinear Case

To contrast the pure linear case, consider now the opposite extreme in which $\beta_2 = 0$. Then (3.1) becomes

$$i\partial A/\partial z = -\gamma |A|^2 A \tag{3.6}$$

and, postulating a solution of the form $A = Ke^{i\varphi}$ and working backwards, it is easy to show that the unique solution to (3.6) is

$$A(z,t) = A(0,t)e^{-i\gamma|A(0,t)|^2 z}. \tag{3.7}$$

Thus the absolute value $|A(z,t)|$ of (3.7), which is the pulse shape, does not change with z. Although it cannot be calculated in closed form, the Fourier transform can be intuitively seen to vary with z because the imaginary exponential in (3.7) has the effect of introducing new frequencies, which causes the transform width to spread. To summarize, in the pure linear case $|\hat{A}(z,\omega)|$ does not change and in the pure nonlinear case $|A(z,t)|$ does not change. These simple calculations go a long way towards interpreting the *dispersion* (β_2) and *nonlinear* (γ) terms in (3.1); roughly speaking, the former causes pulse width variation while the latter causes Fourier transform or spectral variation. They also point out the dual roles of the time and frequency aspects of (3.1).

It must be noted, however, that the association of β_2 with pulse spreading and γ with spectral broadening is actually only partly correct because nonlinearity also indirectly influences pulse width. In fact, if a pulse spectrum is broadened by nonlinearity, then the mismatch in velocities of spectral components becomes larger, thus potentially causing even more temporal pulse broadening. But it should be clear that the derivation of (3.1), which will now be undertaken, must take the Fourier domain into account. The derivation that is provided here is somewhat lengthy, because it contains some details that are missing in other sources, and there are some heuristics. One has the option of reading through the next few pages or accepting (3.1). The end of the derivation is (3.28), which is the spectral domain equivalent of (3.1).

3.1.3 Derivation of NLSE

Taking the curl of the first equation in (2.1), and then using the second equation in (2.1), one has

$$\vec{\nabla}\times(\vec{\nabla}\times\vec{E}) = -\mu_0\left(\vec{\nabla}\times\frac{\partial\vec{H}}{\partial t}\right) = -\mu_0\frac{\partial}{\partial t}(\vec{\nabla}\times\vec{H}) = -\mu_0\frac{\partial^2}{\partial t^2}(\varepsilon_0\vec{E}+\vec{P})$$
$$= -\varepsilon_0\mu_0\frac{\partial^2\vec{E}}{\partial t^2} - \mu_0\frac{\partial^2\vec{P}}{\partial t^2}.$$

Using the familiar identity $\vec{\nabla}\times(\vec{\nabla}\times\vec{E}) = \vec{\nabla}(\vec{\nabla}\bullet\vec{E}) - \Delta^2\vec{E}$ and the last equation in (2.1), one obtains

$$\Delta^2\vec{E} = \varepsilon_0\mu_0\frac{\partial^2\vec{E}}{\partial t^2} + \frac{\mu_0\varepsilon_0}{\varepsilon_0}\frac{\partial^2\vec{P}}{\partial t^2}.$$

Thus the relation $\varepsilon_0\mu_0 = 1/c^2$ (section 2.1) implies

$$\Delta^2\vec{E} - \frac{1}{c^2}\frac{\partial^2\vec{E}}{\partial t^2} = \frac{1}{\varepsilon_0 c^2}\frac{\partial^2\vec{P}}{\partial t^2}, \tag{3.8}$$

where $\Delta^2\vec{E} = (\Delta^2E_x, \Delta^2E_y, \Delta^2E_z)$, $\vec{E} = (E_x, E_y, E_z)$, and Δ^2 is the Laplacian operator $\Delta^2 = \frac{\partial^2}{\partial x^2} + \frac{\partial^2}{\partial y^2} + \frac{\partial^2}{\partial z^2}$. Equation (3.8) is the usual form of the nonlinear wave equation in vector form. There are more general versions based on integral equations and it is possible to waive the assumption $\vec{\nabla}\bullet\vec{E} = 0$ [Men].

As in section 2.1 the polarization vector in (3.8) is assumed to have the form $\vec{P}(x,y,z,t) = \vec{P}_L(x,y,z,t) + \vec{P}_{NL}(x,y,z,t)$, but where the linear and nonlinear terms are now given by the more general expressions [Ag1, NeM, Men, Mil, BuC]

$$\vec{P}_L(x,y,z,t) = \varepsilon_0 \int_{-\infty}^{\infty} \chi^{(1)}(x,y,z,t-\tau)\vec{E}(x,y,z,\tau)d\tau,$$
$$\vec{P}_{NL}(x,y,z,t) = \varepsilon_0 \int_{-\infty}^{\infty} \chi^{(3)}(x,y,z,t-\tau_1,t-\tau_2,t-\tau_3)(\vec{E}(\tau_1)\bullet\vec{E}(\tau_2))\vec{E}(\tau_3)d\vec{\tau}. \tag{3.9}$$

In the second equation of (3.9) the spatial variables in $\vec{E}$ have been suppressed and $d\tau_1 d\tau_2 d\tau_3$ has been written as $d\vec{\tau}$ to shorten the formula. The function $\chi^{(1)}$ is called the *linear susceptibility* and satisfies $\chi^{(1)}(x,y,z,t-\tau) = 0$ for $t < \tau$. The linear susceptibility represents the linear response of the medium to an incident field and takes the form of a weighted memory or delay term. The fact that there can be no response before incidence ($t < \tau$) is known as *causality*. An ideal, instantaneous linear response is modeled by a constant multiple of a delta function, $\chi^{(1)}(s) = const \bullet \delta(s)$, in which case the first of (3.9) becomes $\vec{P}_L = \varepsilon_1 \vec{E}$ as in section 2.1. The *nonlinear susceptibility* term involving $\chi^{(3)}(x,y,z,t_1,t_2,t_3)$ likewise accounts for a delayed nonlinear material response. For general materials there could be, in principle, terms involving $\chi^{(k)}$ for each integer k. However, $\chi^{(2)} = 0$ in silica and the remaining higher order nonlinear terms are typically disregarded [NeM]. Explicit forms for $\chi^{(1)}$ and $\chi^{(3)}$ are provided below; these can be heuristically derived [Ag1]. Expressions (3.9) are the generally accepted forms for the polarization vector $\vec{P} = \vec{P}_L + \vec{P}_{NL}$ in (3.8).

The weakly guiding modal structure discussed in section 2.4 will now be incorporated into (3.8)–(3.9). An overriding assumption for the rest of this book is that the nonlinear term in (3.9) is sufficiently weak, and pulses are sufficiently long, that the basic weakly guiding modal structure derived in section 2.4 continues to hold in the nonlinear setting. Moreover, it is assumed that only the fundamental mode HE_{11} propagates; that is, the fiber under consideration is single mode. In particular, the electric field $\vec{E}$ has just one nonvanishing component (see the discussion below (2.39)). In this case (3.8) and (3.9) reduce to scalar equations for the nonzero component. To simplify notation, the nonzero component will be denoted by E and the corresponding polarization terms by P_L and P_{NL}, respectively, so that (3.8) and (3.9) become

$$\Delta^2 E - \frac{1}{c^2}\frac{\partial^2 E}{\partial t^2} = \frac{1}{\varepsilon_0 c^2}\frac{\partial^2 P}{\partial t^2} \tag{3.10}$$

and

$$P_L(x,y,z,t) = \varepsilon_0 \int_{-\infty}^{\infty} \chi^{(1)}(x,y,z,t-\tau)E(x,y,z,\tau)d\tau,$$
$$P_{NL}(x,y,z,t) = \varepsilon_0 \iiint_{\Re^3} \chi^{(3)}(x,y,z,t-\tau_1,t-\tau_2,t-\tau_3)E(\tau_1)E(\tau_2)E(\tau_3)d\vec{\tau}. \tag{3.11}$$

These are the governing equations for the nonzero component. The triple integral in (3.11) denotes integration over three-dimensional space.

Important further simplification arise from the *slowly varying envelope approximation* (SVEA) [Ag1], which begins by postulating that the signal under consideration has a central carrier frequency ω_0 and that the distributed frequency components are concentrated near ω_0. Real-valued solutions to (3.10) are then postulated to have the form

$$E(x, y, z, t) = \frac{1}{2}T(x, y, z, t)e^{-i\omega_0 t} + cc, \tag{3.12}$$

where cc stands for the complex conjugate. In (3.12) the most rapid oscillations are accounted for by the exponentials; the complex-valued function $T(x, y, z, t)$ is called the *complex envelope* of the carrier ω_0. The first simplification of (3.10) occurs when (3.10) is replaced by an equation for $T(x, y, z, t)$. In fact, if (3.12) is substituted into (3.10) and (3.11), the resulting equation looks as follows:

$$\begin{aligned}
&\left(\Delta^2 - \frac{1}{c^2}\frac{\partial^2}{\partial t^2}\right)\left(\frac{1}{2}Te^{-i\omega_0 t} + cc\right) \\
&= \frac{1}{c^2}\frac{\partial^2}{\partial t^2}\int_{-\infty}^{\infty} \chi^{(1)}\left(\frac{1}{2}Te^{-i\omega_0 \tau} + cc\right)d\tau \\
&+ \frac{1}{c^2}\frac{\partial^2}{\partial t^2}\iiint_{\Re^3} \chi^{(3)}\left(\frac{Te^{-i\omega_0\tau_1} + T^*e^{i\omega_0\tau_1}}{2}\right)\left(\frac{Te^{-i\omega_0\tau_2} + T^*e^{i\omega_0\tau_2}}{2}\right) \\
&\times \left(\frac{Te^{-i\omega_0\tau_3} + T^*e^{i\omega_0\tau_3}}{2}\right)d\vec{\tau}.
\end{aligned} \tag{3.13}$$

The main contributions to the triple integral in (3.13) come when $\tau_1 \cong t$, $\tau_2 \cong t$, $\tau_3 \cong t$, so that there is, roughly speaking, a coefficient of $e^{-i\omega_0 t}$ on the right side of (3.13). Proceeding formally, one may sum up the ways that terms containing $e^{-i\omega_0 t}$ can be obtained on the right side and equate the result to the corresponding expression involving $e^{-i\omega_0 t}$ on the left. There are three ways to obtain the exponential $e^{-i\omega_0 t}$ on the right; specifically, they are the distinct ways to write down the product $(Te^{-i\omega_0 t})(Te^{-i\omega_0 t})(T^*e^{i\omega_0 t})$. Summing, canceling a factor of $1/2$, and equating like exponentials yields

$$\begin{aligned}
\left(\Delta^2 - \frac{1}{c^2}\frac{\partial^2}{\partial t^2}\right)(Te^{-i\omega_0 t}) &= \frac{1}{c^2}\frac{\partial^2}{\partial t^2}\int_{-\infty}^{\infty} \chi^{(1)}Te^{-i\omega_0\tau}d\tau \\
&+ \frac{3}{4c^2}\frac{\partial^2}{\partial t^2}\iiint_{\Re^3} \chi^{(3)}Te^{-i\omega_0\tau_1}Te^{-i\omega_0\tau_2}T^*e^{i\omega_0\tau_3}d\vec{\tau}.
\end{aligned}$$

Introducing $S(x, y, z, t) = T(x, y, z, t)e^{-i\omega_0 t}$ (not the same as (3.12)), the previous equation becomes

$$\begin{aligned}
\Delta^2 S - \frac{1}{c^2}\frac{\partial^2 S}{\partial t^2} &= \frac{1}{c^2}\frac{\partial^2}{\partial t^2}\int_{-\infty}^{\infty} \chi^{(1)}(t-\tau)S(\tau)d\tau \\
&+ \frac{3}{4c^2}\frac{\partial^2}{\partial t^2}\iiint_{\Re^3} \chi^{(3)}(\vec{t} - \vec{\tau})S(\tau_1)S(\tau_2)S^*(\tau_3)d\vec{\tau},
\end{aligned} \tag{3.14}$$

where $\vec{t} - \vec{\tau} = (t - \tau_1, t - \tau_2, t - \tau_3)$. The spatial variables have been suppressed in (3.14) to simplify notation.

Letting $\hat{S}(x, y, z, \omega) = F[S(x, y, z, t)](\omega)$ be the Fourier transform of $S(x, y, z, t)$, (3.14) will now be transformed to the frequency domain. The transform of the left side is

$$\left(\Delta^2 + \frac{\omega^2}{c^2}\right)\hat{S}(x, y, z, \omega). \tag{3.15}$$

Let $\hat{\chi}^{(1)}(x, y, z, \omega) = \int_{-\infty}^{\infty} \chi^{(1)}(x, y, z, t)e^{i\omega\tau}d\tau$ be the transform of the linear susceptibility and note that the linear term on the right of (3.14) is a convolution. Its transform is thus

$$-\frac{\omega^2}{c^2}\hat{\chi}^{(1)}(x, y, z, \omega)\hat{S}(x, y, z, \omega). \tag{3.16}$$

Transferring (3.16) to the left side and adding (3.15) yields

$$(\Delta^2 + \tilde{k}^2(\omega))\hat{S} \tag{3.17}$$

for the total linear part of the transform of (3.14), where

$$\tilde{k}(\omega) = \frac{\omega n(\omega)}{c}, \qquad n(\omega) = (1 + \hat{\chi}^{(1)}(\omega))^{1/2}. \tag{3.18}$$

The term $n(\omega) = \left(1 + \hat{\chi}^{(1)}(\omega)\right)^{1/2}$ in (3.18) is the *general medium index of refraction*. The index is spatially and frequency dependent through the susceptibility term $\hat{\chi}^{(1)}(x, y, z, \omega)$. If the linear response is considered instantaneous and spatially constant, say, in the fiber core, then $n(\omega)$ is constant and $\tilde{k}^2(\omega)$ in (3.17) reduces to the constant n^2k^2 that was used in (2.28).

Turning to the nonlinear part of (3.14), define

$$\hat{\chi}^{(3)}(x, y, z, u, v, \omega) = \iiint_{\Re^3} \chi^{(3)}(x, y, z, \vec{\tau})e^{iu\tau_1}e^{iv\tau_2}e^{i\omega\tau_3}d\vec{\tau},$$

the three-dimensional Fourier transform. It is straightforward but tedious to show that the nonlinear term in (3.14) then transforms to

$$-\frac{3\omega^2}{16\pi^2c^2}\iint_{\Re^2} \hat{\chi}^{(3)}(u, v-u, \omega-v)\hat{S}(u)\hat{S}(v-u)\hat{S}^*(v-\omega)du\,dv, \tag{3.19}$$

where the Fourier transform relationship $F[S^*(t)](\omega) = \hat{S}^*(-\omega)$ and the fact that the nonlinear term in (3.14) is a convolution expression have been used. Equating (3.17) and (3.19) one has

$$\begin{aligned}(\Delta^2 + \tilde{k}^2(\omega))\hat{S}(\omega) \\ = -\frac{3\omega^2}{16\pi^2c^2}\iint_{\Re^2} \hat{\chi}^{(3)}(u, v-u, \omega-v)\hat{S}(u)\hat{S}(v-u)\hat{S}^*(v-\omega)du\,dv,\end{aligned} \tag{3.20}$$

where the spatial variables have again been suppressed.

It will be useful to separate out the transverse and longitudinal variables in (3.20) by postulating $\hat{S}(x, y, z, \omega) = F(x, y, \omega)\hat{U}(z, \omega)e^{i\beta(\omega)z}$. Substituting, the left side of (3.20) becomes

$$(\Delta_T^2 F)\hat{U}e^{i\beta(\omega)z} + (\hat{U}_{zz} + 2i\beta(\omega)\hat{U}_z - \beta^2(\omega)\hat{U})Fe^{i\beta(\omega)z} + \tilde{k}^2(\omega)F\hat{U}e^{i\beta(\omega)z}, \tag{3.21}$$

where Δ_T^2 denotes the transverse Laplacian (x and y variables) and the z subscript denotes a derivative. In the SVEA $\hat{U}_{zz}$ is assumed to be small in comparison to the other terms in (3.21) [Ag1] and is consequently dropped; justification for the SVEA will be discussed briefly at the end of this section. Dropping $\hat{U}_{zz}$ leaves (3.21) in the form

$$(\Delta_T^2 F + (\tilde{k}^2(\omega) - \beta^2(\omega))F)\hat{U}e^{i\beta(\omega)z} + 2i\beta(\omega)\hat{U}_z F e^{i\beta(\omega)z}. \tag{3.22}$$

Set the first term in (3.22) to 0, $\Delta_T^2 F + (\tilde{k}^2(\omega) - \beta^2(\omega))F = 0$, impose the weakly guiding conditions on $F(x, y, \omega)$ at the core–cladding interface for each ω, and choose $\beta(\omega)$ to be the propagation constant corresponding to the fundamental mode. Thus $F(x, y, \omega)$ and $\beta(\omega)$ are chosen to satisfy the scalar modal equation (2.28) for each ω. Then (3.22) reduces to simply the last term, and consequently (3.20) becomes

$$\begin{aligned}
&2i\beta(\omega)\,\hat{U}_z F \\
&= -\frac{3\omega^2}{16\pi^2 c^2}\iint_{\Re^2} \hat{\chi}^{(3)} F(u)F(v-u)F^*(v-\omega)\hat{U}(u)\hat{U}(v-u)\hat{U}^*(v-\omega)e^{i\Delta\beta(\omega)z}du\,dv,
\end{aligned} \tag{3.23}$$

where $\Delta\beta(\omega) = \beta(u) + \beta(v-u) - \beta(v-\omega) - \beta(\omega)$ and the spatial variables have been suppressed. This is the general form of the slowly varying NLSE in the spectral domain. What follows now is a series of simplifications that will reduce (3.23) to a manageable form.

Using an approach of [BIW] one can remove the transverse dependence in (3.23) by confining (x, y) to the fiber core, where the nonlinear susceptibility $\hat{\chi}^{(3)}$ is assumed constant as a function of (x, y), and then integrating with respect to x and y. Introducing

$$\begin{aligned}
R_1(\omega) &= \iint_{\Re^2} F(x, y, \omega)dx\,dy, \\
R_2(u, v, \omega) &= \iint_{\Re^2} F(x, y, u)F(x, y, v)F^*(x, y, \omega)dx\,dy,
\end{aligned}$$

(3.23) becomes

$$\begin{aligned}
&\hat{U}_z R_1(\omega) \\
&= -\frac{3\omega^2}{32i\pi^2\beta(\omega)c^2}\iint_{\Re^2} \hat{\chi}^{(3)} R_2(u, v-u, v-w)\hat{U}(u)\hat{U}(v-u)\hat{U}^*(v-\omega)e^{i\Delta\beta(\omega)z}du\,dv.
\end{aligned} \tag{3.24}$$

It is still assumed that $n_2^2 k(\omega)^2 \le \beta(\omega)^2 \le n_1^2 k(\omega)^2$ (see below (2.12)), although $\beta(\omega)$ and $k(\omega)$ are now frequency dependent. Then $n_2^2 \le \frac{\beta(\omega)^2}{k(\omega)^2} \le n_1^2$, so that $\frac{1}{n_1^2} \le \frac{\omega^2}{c^2\beta(\omega)^2} \le \frac{1}{n_2^2}$. In the weakly guiding case $n_1 \cong n_2$ and so the coefficient of the integral in (3.24) is thus approximately constant. Also, the functions R_1 and R_2 are weakly dependent on frequency and are usually set to constants [Ag1]. Making these substitutions, (3.24) reduces to

$$\hat{U}_z = i\kappa\iint_{\Re^2} \hat{\chi}^{(3)}\hat{U}(u)\hat{U}(v-u)\hat{U}^*(v-\omega)e^{i\Delta\beta(\omega)z}du\,dv, \tag{3.25}$$

where κ is a constant and (3.25) depends only on z and ω.

There are several suggested forms for the nonlinear susceptibility $\chi^{(3)}$ [Ag1, BIW], but in the simplest approximation it is taken to be a constant multiple of a delta function,

$\chi^{(3)}(x, y, z, t_1, t_2, t_3) = \text{constant} \bullet \delta(t_1, t_2, t_3)$, in which case the transform $\hat{\chi}^{(3)}$ is a constant and (3.25) becomes

$$\hat{U}_z(z, \omega) = i\kappa' \iint_{\Re^2} \hat{U}(z, u)\hat{U}(z, v-u)\hat{U}^*(z, v-\omega)e^{i\Delta\beta(\omega)z}du\,dv \tag{3.26}$$

for another constant κ'.

The exponentials can be removed from (3.26) by appeal to the power series expansion of $\beta(\omega)$, which has the form

$$\beta(\omega) = \beta_0 + \beta_1(\omega - \omega_0) + (\beta_2/2)(\omega - \omega_0)^2 + \cdots, \tag{3.27}$$

where ω_0 is the carrier. The term $\Delta\beta(\omega) = \beta(u) + \beta(v-u) - \beta(v-\omega) - \beta(\omega)$ simplifies to

$$\Delta\beta(\omega) = (\beta_2/2)[(u-\omega_0)^2 + (v-u-\omega_0)^2 - (v-\omega-\omega_0)^2 - (\omega-\omega_0)^2]$$

because of cancellation of the constant and linear terms. Certainly higher order coefficients $(\beta_3, \beta_4, \dots)$ could be used in (3.27), but for now only the quadratic coefficient will be kept. The last step is to let

$$\hat{A}(z, \omega - \omega_0) = \hat{U}(z, \omega)e^{i(\beta_2/2)(\omega-\omega_0)^2 z}$$

and substitute into (3.26) to obtain

$$\hat{A}_z(z, \omega) - i\frac{\beta_2}{2}\omega^2\hat{A}(z, \omega) = i\kappa' \iint_{\Re^2} \hat{A}(z, u)\hat{A}(z, v-u)\hat{A}^*(z, v-\omega)du\,dv \tag{3.28}$$

($\kappa' = \text{constant}$), where the frequency variable has been shifted. The inverse Fourier transform of (3.28) is

$$\frac{\partial A}{\partial z} + i(\beta_2/2)\frac{\partial^2 A}{\partial t^2} = i\gamma|A|^2 A, \tag{3.29}$$

where γ is another constant. Since (3.29) is the same as (3.1), this completes the derivation of the NLSE.

The constant γ in (3.29) can be determined experimentally by expressing it in terms of the *nonlinear index* n_2 [Ag1] (not to be confused with the cladding index of refraction, which uses the same symbol), defined by $n_2 = \gamma c A_{eff}/\omega_0$, where c is the vacuum speed of light, ω_0 is the carrier frequency, and A_{eff} is the *effective area* of the fiber. The nonlinear index of a fiber can be measured [Ag1, App. B]. The constant γ is called the *Kerr* coefficient; the *Kerr effect* is the term used to describe the dependence of the index of refraction on light intensity.

Note that (3.1) is actually the equation for the envelope (slowly varying) of the electric field E. Knowing the propagation constant $\beta(\omega)$, which can be found from the scalar wave equation (2.28), one obtains E by back-substitution through the changes of variable leading to (3.29).

To briefly motivate the SVEA used in (3.22), start with $n_2^2 \le \frac{\beta(\omega)^2}{k(\omega)^2} \le n_1^2$, $k(\omega) = \omega/c$, as below (3.24). Then $\beta(\omega) = O(k(\omega))$, where the symbol $O(*)$ stands for *order of* *magnitude.* Now for $\lambda = 1.55\mu m$, one has $\omega = O(10^{15}s^{-1})$, and thus $k(\omega)$ has order $O(10^{15}s^{-1}/10^{8}ms^{-1}) = O(10^{10}km^{-1})$, which means that $\beta(\omega)$ has order $10^{10}km^{-1}$. Recalling that $\hat{U}$ is dimensionless (see the discussion below (3.1)), the terms in (3.21) containing $\beta(\omega)$ are dominant and this justifies dropping $\hat{U}_{zz}$. For an even simpler motivation, in the linear case where $\vec{P}_L = \varepsilon_1 \vec{E}$ and $\vec{P}_{NL} = 0$, (3.10) becomes $\Delta^2 E = \frac{1+(\varepsilon_1/\varepsilon_0)}{c^2}\frac{\partial^2 E}{\partial t^2}$. If the transverse dependence is constant, $\partial_x = \partial_y = 0$, then this reduces to $E_{zz} = C^2 E_{tt}$ with $C = O(1/c)$. In the conventional units used here one can already see that E_{zz} is much smaller than E_{tt}.

It is instructive to check (3.28) and (3.29) for consistency with the linear analysis discussed at the end of Chapter 2. If the nonlinearity is negligible, then $\kappa' = 0$ in (3.26) and so $\hat{U}(z,\omega)$ is constant in z, where now $\hat{S}(x,y,z,\omega) = F(x,y,\omega)\hat{U}(\omega)e^{i\beta(\omega)z}$. If the spectrum is narrow, $\omega \cong \omega_0$, as typical, then $F(x,y,\omega)$ and $\beta(\omega)$ could be viewed as roughly independent of ω; in other words, $\hat{S} = F(x,y)\hat{U}(\omega)e^{i\beta z}$ and, in terms of $\hat{T}(\omega) = \hat{S}(\omega + \omega_0)$ (see above (3.14)), $\hat{T}(\omega) = F(x,y)\hat{U}(\omega+\omega_0)e^{i\beta z}$. Letting $f(t)$ be the inverse transform of $\hat{U}(\omega+\omega_0)$, one has $T(t) = F(x,y)f(t)e^{i\beta z}$, and from (3.12), $E = (1/2)F(x,y)f(t)e^{-i(\omega t - \beta z)} + cc$. This is consistent with the linear form (2.40), recalling that there is only one nonzero component.

3.2 Interpretations of the NLSE

Going back to the form of $\Delta\beta$ below (3.27), note that the linear terms $\beta_1(w - \omega_0)$ could have been retained, instead of canceled, and $\Delta\beta$ would have been

$$\begin{aligned}\Delta\beta(\omega) = (\beta_1)[(u-\omega_0) + (v-u-\omega_0) - (v-\omega-\omega_0) - (\omega-\omega_0)] \\ + (\beta_2/2)[(u-\omega_0)^2 + (v-u-\omega_0)^2 - (v-\omega-\omega_0)^2 - (\omega-\omega_0)^2].\end{aligned}$$

Carrying on with the derivation with the new $\Delta\beta$ would lead to

$$i\frac{\partial A}{\partial z} = -i\beta_1\frac{\partial A}{\partial t} + (\beta_2/2)\frac{\partial^2 A}{\partial t^2} - \gamma|A|^2 A \tag{3.30}$$

instead of (3.1). It turns out that the time variable in (3.30) is physical time, whereas that in (3.1) is local time [Ag1], and making the connection gives an interpretation of the β_1 constant. For purposes of interpretation, ignore the terms involving β_2 and γ for the moment, leaving (3.30) in the form $A_z + \beta_1 A_t = 0$. Letting $T = T(t,z) = t - \beta_1 z$ and defining $B(z,T) = A(z,t) = A(z, T + \beta_1 z)$, the previous equation $A_z + \beta_1 A_t = 0$ is equivalent to $B_z = 0$ so that $B(z,T) = B(T)$, a function of T only. That is, the solution has the form $A(z,t) = B(t - \beta_1 z)$ for some function B. This is a traveling wave in the fixed shape $B(t)$ with wave speed $v_g = 1/\beta_1$. The term v_g is the *group velocity* mentioned at the beginning of Chapter 2 and attaches a single primary wave speed to an entire optical pulse. Dispersion as represented by β_2 can be considered to be a perturbation of the group velocity whereby certain portions of a pulse may travel faster or slower than others relative to the group velocity. Since $\beta_2 = \beta''(\omega_0)$ and $\beta_1 = \beta'(\omega_0)$, and β_1 is associated with group velocity, β_2 is sometimes called the *group velocity dispersion*.

If one uses the same substitution $T = T(t,z) = t - \beta_1 z$ in (3.30), then the equation

(3.1) is obtained with T in place of t, that is,

$$i\frac{\partial A}{\partial z} = (\beta_2/2)\frac{\partial^2 A}{\partial T^2} - \gamma|A|^2 A. \tag{3.31}$$

Thus T is the local time variable set to 0 relative to the group velocity. To keep notation as simple as possible, (3.1) will be the reference NLSE but the symbol t will be regarded as local time, or the T in (3.31).

3.2.1 Perturbation Series

An interpretation of the nonlinear term in (3.28) can be given by an elementary perturbation argument. The factor $i\kappa'$ in (3.28) is typically small, and this suggests setting $\varepsilon = i\kappa'$ and expanding in a regular perturbation series. To this end let

$$\hat{A}(z,\omega) = \hat{A}^{(0)}(z,\omega) + \varepsilon\hat{A}^{(1)}(z,\omega) + \varepsilon^2\hat{A}^{(2)}(z,\omega) + \cdots$$

and substitute into (3.28) to obtain a system of equations by equating powers of ε in the standard way. The first two coefficients are (suppressing limits of integration)

$$\begin{aligned}
\varepsilon^0:&\quad \hat{A}^{(0)}_z - (i\beta_2\omega^2/2)\hat{A}^{(0)}(z,\omega) = 0,\\
\varepsilon^1:&\quad \hat{A}^{(1)}_z - (i\beta_2\omega^2/2)\hat{A}^{(1)}(z,\omega) = \iint \hat{A}^{(0)}(z,u)\hat{A}^{(0)}(z,v-u)\hat{A}^{(0)*}(z,v-\omega)\,du\,dv,\\
&\quad \hat{A}^{(0)}(0,\omega) = \hat{A}(0,\omega), \qquad \hat{A}^{(k)}(0,\omega) = 0, \quad k > 0,
\end{aligned} \tag{3.32}$$

the solution to the first of which is $\hat{A}^{(0)}(z,\omega) = \hat{A}(0,\omega)e^{i\beta_2\omega^2 z/2}$. Substituting this into the second of (3.32) and multiplying by the integrating factor $e^{-i\beta_2\omega^2 z/2}$ gives

$$(\hat{A}^{(1)}e^{-i\beta_2\omega^2 z/2})_z = \iint \hat{A}(0,u)\hat{A}(0,v-u)\hat{A}^*(0,v-\omega)e^{i\tilde{\Delta}\beta_2 z}\,du\,dv, \tag{3.33}$$

where $\tilde{\Delta}\beta_2 = (\beta_2/2)[u^2 + (v-u)^2 - (v-\omega)^2 - \omega^2]$. Integrating (3.33) and using the initial conditions in (3.32) gives

$$\hat{A}^{(1)}e^{-i\beta_2\omega^2 z/2} = \iint \hat{A}(0,u)\hat{A}(0,v-u)\hat{A}^*(0,v-\omega)\frac{e^{i\tilde{\Delta}\beta_2 z} - 1}{i\tilde{\Delta}\beta_2}\,du\,dv.$$

When $\tilde{\Delta}\beta_2 z$ is small the last expression is approximately

$$\hat{A}^{(1)}e^{-i\beta_2\omega^2 z/2} \cong z\iint \hat{A}(0,u)\hat{A}(0,v-u)\hat{A}^*(0,v-\omega)\,du\,dv.$$

The perturbation series through the linear term thus looks like

$$\hat{A}(z,\omega) \cong e^{i\beta_2\omega^2 z/2}\left[\hat{A}(0,\omega) + \varepsilon z\iint \hat{A}(0,u)\hat{A}(0,v-u)\hat{A}^*(0,v-\omega)\,du\,dv\right]. \tag{3.34}$$

The first term on the right of (3.34) can be viewed as the linear approximation to the solution of (3.1); in fact, it agrees exactly with (3.2) in the linear case. Hence every solution is, over sufficiently small propagation length, the linear term plus a perturbation term, and it is the latter that is of interest in the spectral domain of (3.34). The width of the modulus on the left side of (3.34) is the spectral width. The width of the first term in (3.34), that is, the linear term, is the same as the initial width in (3.2). However, the term

$$\iint \hat{A}(0,u)\hat{A}(0,v-u)\hat{A}^*(0,v-\omega)du\,dv$$

is a double convolution that is (intuitively at least) broader than the individual term $\hat{A}(0,\omega)$. As z increases from $z = 0$, (3.34) has the effect of growing in new frequency components of the pulse $A(z,t)$.

3.2.2 Frequency Generation

Some insight into how new frequencies are generated can be found by going again to (3.28). Imagine that the initial transform $\hat{A}(0,\omega)$ has significant frequency components at ω_1, ω_2, and $2\omega_2-\omega_1$, where ω_1 and ω_2 are near each other, but no component at $2\omega_1-\omega_2$. To take a trivial example, let $\omega_1 = 99$ and $\omega_2 = 101$, so that $2\omega_2-\omega_1 = 103$ and $2\omega_1-\omega_2 = 97$ (the units do not matter for purposes of illustration). Thus there are initially components at 99, 101, and 103, but none at 97. In (3.28) let $u \cong \omega_1$, $v \cong \omega_1+\omega_2$, and $\omega \cong 2\omega_1-\omega_2 = 97$. In neighborhoods of $u \cong \omega_1$, $v-u \cong \omega_2$, and $v-\omega = 2\omega_2-\omega_1$, all terms under the integral in (3.28) are nonzero, and so intuitively the left side becomes nonzero for z sufficiently large. In other words, a component at $\omega \cong 97$ grows in because of (3.28). By switching roles of the variables, one sees that a component also appears at 105. In this way the nonlinear term in (3.14) produces new frequencies that appear in the form of a broadened spectrum.

3.3 The Linear Gaussian Model

The NLSE can be solved in closed form in very few cases. The pure linear and pure nonlinear scenarios, where analytical solutions are available, were encountered in section 3.1. The important case of solitons will be covered in Chapter 5. But here the pure linear case involving Gaussian initial pulses will be discussed in detail because it serves as a guide to the general theory of pulse behavior. One reason for the importance is the familiar property that the Fourier transform of a Gaussian shape is another Gaussian shape. Another key reason is that, in the pure linear case, a pulse that is initially Gaussian remains Gaussian during propagation.

To see this, consider (3.1) with $\gamma = 0$ and proceed via (3.2) to

$$\hat{A}(z,\omega) = \hat{A}(0,\omega)e^{i\beta_2\omega^2 z/2},$$

where $\hat{A}(0,\omega)$ is the transform of the initial pulse shape $A(0,t)$. By the inverse transform,

$$A(z,t) = \frac{1}{2\pi}\int_{-\infty}^{\infty} \hat{A}(0,\omega)e^{i\beta_2\omega^2 z/2}e^{-i\omega t}d\omega \tag{3.35}$$

for all z. Now suppose that $A(0,t)$ is Gaussian, say,

$$A(0,t) = Me^{-\frac{t^2}{2T_0^2}}, \tag{3.36}$$

where M and T_0 are positive constants; here T_0 is the standard half-width–half-maximum (HWHM) width index (see Chapter 1). The Fourier transform relationship between Gaussians is

$$\int_{-\infty}^{\infty} e^{-\frac{\tau^2}{2\Gamma}} e^{i\omega\tau} d\tau = \sqrt{2\pi\Gamma} e^{-\frac{\Gamma\omega^2}{2}}, \tag{3.37}$$

and so using $\Gamma = T_0^2$ the transform of (3.36) is $\hat{A}(0,\omega) = M\sqrt{2\pi T_0^2} e^{-\frac{T_0^2\omega^2}{2}}$. Feeding this into (3.35) gives

$$\begin{aligned} A(z,t) &= \frac{1}{2\pi} M T_0 \sqrt{2\pi} \int_{-\infty}^{\infty} e^{-\frac{T_0^2\omega^2}{2}} e^{i\beta_2\omega^2 z/2} e^{-i\omega t} d\omega \\ &= \frac{MT_0}{\sqrt{T_0^2 - i\beta_2 z}} \frac{1}{2\pi} \int_{-\infty}^{\infty} \sqrt{2\pi(T_0^2 - i\beta_2 z)} e^{-\frac{(T_0^2 - i\beta_2 z)}{2}\omega^2} e^{-i\omega t} d\omega \\ &= \frac{MT_0}{\sqrt{T_0^2 - i\beta_2 z}} e^{-\frac{t^2}{2(T_0^2 - i\beta_2 z)}}, \end{aligned} \tag{3.38}$$

where in the last step $\Gamma = T_0^2 - i\beta_2 z$ in (3.37). As a function of t, the pulse envelope $A(z,t)$ is thus Gaussian. Note that the amplitude of the central peak ($t = 0$) decreases as $1/\sqrt{z}$ with increasing propagation distance z.

Written in more suggestive form, (3.38) is

$$A(z,t) = \frac{MT_0\sqrt{T_0^2 + i\beta_2 z}}{T_0^4 + \beta_2^2 z^2} e^{-\frac{t^2}{2(T_0^2 + (\beta_2^2 z^2/T_0^2))}} e^{-i\frac{\beta_2 z t^2}{T_0^4 + \beta_2^2 z^2}}, \tag{3.39}$$

a formula that provides the modulus and phase of the envelope in explicit form. In particular, note that a real-valued initial envelope $A(0,t)$ becomes complex valued and that its time varying phase, given by the second exponential in (3.39), is quadratic in t. Now imagine that a real-valued pulse is launched into a fiber with a certain dispersion constant β_2. After propagating for some distance the fiber is spliced into a second fiber with a different dispersion constant $\tilde{\beta}_2$. For purposes of the second fiber, the pulse is initially a Gaussian shape with quadratically varying phase (3.39), which was acquired from propagating in the first fiber. This suggests that more general initial conditions than (3.36) might be considered and that a reasonable candidate is

$$A(0,t) = Me^{-\frac{t^2}{2T_0^2}(1+iC)}, \tag{3.40}$$

where C is a constant. The constant C in (3.40) is called the initial *chirp*, and (3.40) is referred to as a *chirped Gaussian* input pulse. The term "chirp" seems to have come from radar and refers to a change in frequency or pitch across the pulse. That is, for example, the front and back edges of the pulse, in the time variable, have different local frequencies, or dominant frequencies of a small sample of the pulse. It turns out that appropriately chirped pulses can have desirable propagation features and these will be considered in the next chapter. Now, however, it will be noted that an initial Gaussian pulse, with chirp, remains Gaussian during propagation; the proof is only a little more complicated than (3.38). Starting with (3.40) and emulating the above steps shows that the output pulse envelope at distance z is

$$A(z,t) = \frac{MT_0}{\sqrt{T_0^2 - i\beta_2 z(1+iC)}} e^{-\frac{(1+iC)t^2}{2(T_0^2 - i\beta_2 z(1+iC))}}. \tag{3.41}$$

The exponential in (3.41) can be written

$$\frac{-t^2}{2\left[\left(T_0 + \frac{\beta_2 C z}{T_0}\right)^2 + \left(\frac{\beta_2 z}{T_0}\right)^2\right]} + i(*), \tag{3.42}$$

where the asterisk denotes the appropriate imaginary part, not of interest for the present. The real part of (3.42) gives the HWHM of the output amplitude $|A(z,t)|$ as

$$T_{out}^2 = \left(T_0 + \frac{\beta_2 C z}{T_0}\right)^2 + \left(\frac{\beta_2 z}{T_0}\right)^2 = T_0^2 + 2\beta_2 C z + \frac{\beta_2^2 z^2}{T_0^2}(1+C^2). \tag{3.43}$$

Note first that (3.43) gives the formula (1.1) when $C = 0$. Furthermore, a number of interesting conclusions follow from (3.43).

For example, suppose that C, β_2, and z are fixed and, in an elementary pulse design exercise, one wishes to minimize the output pulse width (3.43) by appropriately choosing the input width T_0. The graph of (3.43), as a function of T_0^2, is nonnegative and unbounded as either $T_0 \to 0$ or $T_0 \to \infty$, indicating an absolute minimum of T_{out}. Differentiation of (3.43) with respect to T_0^2 and setting the derivative to 0 gives the critical point as $T_{0,crit}^2 = \beta_2 z\sqrt{1+C^2}$ and a corresponding minimum in (3.43) of $T_{out,\min}^2 = 2\beta_2 z(C + \sqrt{1+C^2})$. Thus, while the pulse width T_{out} in (3.43) expands linearly with distance, the *optimal width* grows by order of magnitude $\sqrt{z}$.

What is it about dispersion that causes the output width to become larger if the input width is made smaller? Some insight can be provided by looking at the Fourier transform of (3.40), which is (take $\Gamma = T_0^2/(1+iC)$ in (3.37))

$$\hat{A}(0,\omega) = MT_0\sqrt{2\pi/(1+iC)}e^{-\frac{T_0^2(1-iC)}{2(1+C^2)}\omega^2}. \tag{3.44}$$

The HWHM of (3.44) is $W_0^2 = (1+C^2)/T_0^2$, and so (3.44) implies the fundamental relationship [Ag1]

$$T_0^2 W_0^2 = \sqrt{1+C^2} \tag{3.45}$$

between the temporal and spectral widths. Thus making T_0^2 smaller simply forces W_0^2 to be larger. That in turn means that more frequencies are present, which leads to wider disparity among wave speeds of pulse components, and thus further overall dispersion. This and the pulse design illustration of the previous paragraph are the kinds of simple and insightful analytical calculations that, as will be seen in the next chapter, provide a pattern in the general nonlinear picture.

3.4 Other Considerations

This section will round out the chapter on governing equations by indicating other terms that may be included in (3.1), the role of the magnetic field $\vec{H}$ (which has not been mentioned so far in connection with the NLSE), alternate forms of (3.1), and a few closing remarks.

Perhaps the most important omission from (3.1) is a term corresponding to energy loss, or signal attenuation. It is instructive to note first that solutions of (3.1) conserve energy. As mentioned in Chapter 1, the total energy in the pulse is

$$E = \int_{-\infty}^{\infty} |A(z,t)|^2 dt,$$

whose derivative with respect to z is

$$\frac{dE}{dz} = \int_{-\infty}^{\infty} [A(z,t)A(z,t)^*]_z dt = \int_{-\infty}^{\infty} [AA_z^* + A_z A^*] dt.$$

Substituting from (3.1) and integrating by parts shows that $dE/dz = 0$; that is, conservation of energy holds for (3.1). Since energy is certainly not conserved, there should be an attenuation factor included in (3.1), and the appropriate one is a term of the form $-i\alpha A(z,t)$ on the right side [Ag1]. Such an expression arises naturally in the derivation of the NLSE when the propagation constant $\beta(\omega)$ is permitted to be complex valued, where the imaginary part assumes the role of loss [Ag1, NeM]. However, even without loss it is possible to interpret solutions of (3.1) as physically realistic. If the constant γ is interpreted as a certain average over the loss and amplification cycle, then (3.1) remains approximately valid [BPJ]. This will be discussed again in the context of the dimensionless form of (3.1) later in this section (see the paragraph below (3.50)).

As a mathematical comment, note that the conservation condition

$$E = \int_{-\infty}^{\infty} |A(z,t)|^2 dt = \int_{-\infty}^{\infty} |A(0,t)|^2 dt = \text{constant}$$

shows that solutions of (3.1) are uniformly bounded in the L^2 norm if the initial energy is finite.

More terms can also be added to (3.1) if they are retained in the series expansion (3.27), for example, the term containing β_3 (see the next equation below). Without going into details about its derivation, the most general form of the NLSE typically used is [Ag1]

$$\frac{\partial A}{\partial z} + \alpha A + i(\beta_2/2)\frac{\partial^2 A}{\partial t^2} - (\beta_3/6)\frac{\partial^3 A}{\partial t^3} = i\gamma\left(|A|^2 A + (i/\omega_0)\frac{\partial(|A|^2 A)}{\partial t} - T_R A\frac{\partial |A|^2}{\partial t}\right), \tag{3.46}$$

where the new terms on the second line of (3.46) correspond to the self-steepening or shock effect and the delayed Raman response, respectively [Ag1].

In connection with the magnetic field $\vec{H}$, note that the discussion in this chapter has focused on the envelope $A(z, t)$ of the scalar electric field $E(z, t)$. Going back to (2.23) and recalling the weakly guiding assumption that only one component of $\vec{E}$ is nonzero, namely, E_y, it follows that only H_x is nonzero in $\vec{H}$. In fact, H_x is given explicitly in terms of E_y by (2.23). Actually, (2.23) gives two relationships, namely, the second and fourth equations, which reduce to $i\beta H_x = -i\omega\varepsilon E_y$ and $-i\beta E_y = i\omega\mu_0 H_x$. These equations are compatible only if $\beta^2 = \omega^2\mu_0\varepsilon$. However, in the fiber core region one has $\omega^2\mu_0\varepsilon = \omega^2\mu_0\varepsilon_0(\varepsilon/\varepsilon_0) = (\omega^2/c^2)n_1^2 = k^2n_1^2$ by (2.2) and the definition of the wave number k defined above (2.11). The argument below (3.24), concerning the coefficient of the integral in that equation, shows that β^2 is squeezed between $k^2n_1^2$ and $k^2n_2^2$ in the weakly guiding assumption $n_1 \cong n_2$. Thus $\beta^2 \cong \omega^2\mu_0\varepsilon$, so the two relationships connecting H_x and E_y are compatible (approximately). Hence $\vec{H}$ is determined as essentially a constant multiple of $\vec{E}$, which in turn is determined by $A(z, t)$. For this reason it is customary in fiber optics to study only the envelope $A(z, t)$ instead of the total field $(\vec{E}, \vec{H})$ (although it is not usually explained why it is permissible to do so).

It is often convenient to write (3.1) in normalized dimensionless units. This is done conventionally by defining the peak power P_0 by $P_0 = |A(0, 0)|^2$ and then

$$U(z, t) = \frac{A(z, t)}{\sqrt{P_0}}.$$

The symbol $U(z, t)$ here does not have the same meaning as that defined in (3.21) and is used in order to be consistent with [Ag1]. The equation for $U(z, t)$, which is dimensionless, is

$$i\frac{\partial U}{\partial z} = (\beta_2/2)\frac{\partial^2 U}{\partial t^2} - (\gamma P_0)|U|^2U, \quad |U(0, 0)| = 1. \tag{3.47}$$

Now select some reference pulse width T_0, such as the HWHM or RMS width, and let $\tau = t/T_0$ so that $T_0^2 U_{tt} = U_{\tau\tau}$. Note that τ is a dimensionless time variable. Then (3.47) is the same as

$$i\frac{\partial U}{\partial z} = \frac{\beta_2}{2T_0^2}\frac{\partial^2 U}{\partial \tau^2} - (\gamma P_0)|U|^2U. \tag{3.48}$$

In (3.48) β_2 has units of ps^2/Km and thus β_2/T_0^2 has units of inverse length. Define $L_D = T_0^2/|\beta_2|$, a quantity called the *dispersion length*, and introduce a dimensionless length by $\zeta = z/L_D$. The dispersion length is a benchmark at which the squared output width of an unchirped ($C = 0$) Gaussian pulse doubles; this can be seen by substituting $z = L_D = T_0^2/|\beta_2|$ into (3.43). Inserting $\zeta = z/L_D$ into (3.48) gives

$$\frac{i}{L_D}\frac{\partial U}{\partial \zeta} = \frac{sign(\beta_2)}{2L_D}\frac{\partial^2 U}{\partial \tau^2} - (\gamma P_0)|U|^2U. \tag{3.49}$$

The constant γ has units of inverse length times inverse power, that is, $mW^{-1}Km^{-1}$, so that γP_0 has units of inverse length. In fact, $L_{NL} = (\gamma P_0)^{-1}$ is called the *nonlinear length*.

Multiplying (3.49) by L_D gives

$$i\frac{\partial U}{\partial \zeta} = \frac{sign(\beta_2)}{2}\frac{\partial^2 U}{\partial \tau^2} - N^2|U|^2U, \tag{3.50}$$

where N^2 is defined by $N^2 = L_D/L_{NL}$. The constant N is called the *normalized nonlinearity constant* and measures the departure of (3.50) from the pure linear case $\gamma = 0$ considered in section 3.1. The equation (3.50) is entirely dimensionless (both independent and dependent variables) and is a standard form of the NLSE [Ag1].

In the average power approach mentioned earlier in connection with fiber attenuation, the constant P_0 is taken to be the average power over a span of fiber with an amplifier at the end. In this ensemble approach the loss term can be disregarded and (3.1) is acceptably accurate [BPJ]. Another approach using average power is discussed at the end of section 4.2.

The constant N can be removed by making $u(\zeta, \tau) = NU(\zeta, \tau)$ so that $|u(0, 0)| = N$ and $iu_\zeta = (sign(\beta_2/2))u_{\tau\tau} - |u|^2u$. In the soliton literature, which will be taken up in Chapter 5, the case $sign(\beta_2) < 0$ is called the *focusing case* and $sign(\beta_2) > 0$ is called the *defocusing case*:

$$\begin{aligned} &iu_\zeta + (1/2)u_{\tau\tau} + |u|^2u = 0, \quad \text{focusing,} \\ &iu_\zeta - (1/2)u_{\tau\tau} + |u|^2u = 0, \quad \text{defocusing.} \end{aligned} \tag{3.51}$$

Solitons can form in the focusing case but not otherwise. In the optical communications literature the case $sign(\beta_2) > 0$ is also called the *normal dispersion* case, while $sign(\beta_2) < 0$ is termed the *anomalous dispersion* case [Ag1]:

$$\begin{aligned} &\text{normal dispersion,} \quad && sign(\beta_2) > 0, \\ &\text{anomalous dispersion,} \quad && sign(\beta_2) < 0. \end{aligned}$$

Hence solitons are supported only in the anomalous regime. This will be discussed in detail in both Chapters 4 and 5.

Aside from the pure linear, pure nonlinear, and soliton cases, (3.1) is not solvable in closed form [Ag1, AbS] and rigorous analysis and simulation can only be performed numerically. The standard procedure for numerically solving (3.1) is the *split step Fourier method* [Ag1], a technique that is interesting in its own right and fits in appropriately with the topics in this chapter. In the split step method, dispersion and nonlinearity are imagined to act independently and sequentially over small time steps. If h is a small increment in the z variable in (3.1), one imagines that β_2 can be set to 0 between $z = 0$ and $z = h/2$, then $\gamma = 0$ from $z = h/2$ to $z = h$. Thus nonlinearity acts alone in the first subinterval, and dispersion alone in the second. On each subinterval the equation is, respectively, pure nonlinear and pure linear and can thus be solved as in section 3.1. The pure linear solution is found by using the fast Fourier transform. The process is iterated until the desired endpoint $z = L$ is reached. A *modified split step* procedure [Ag1] is an order of magnitude faster than the finite element method and is accurate to order $O(h^2)$. The split step method is the industry standard, has been proved accurate by countless trials, and is considered to be essentially exact [Ag1].

To finish out the section and chapter it should be mentioned that, although the discussion here has taken place in the context of a single carrier frequency and single isolated pulse, the optics community is also interested in multiple carriers (or channels) and sequences of pulses. The NLSE remains valid for multiple pulses, but coupled equations need to be used for multichannel modeling [Ag1]. It should also be mentioned that optical noise is a critical impairment in long-distance fiber systems; optical amplifiers enhance both the noise and the signal. Noise effects are largely statistical and are not considered here.

Chapter 4

The Variational Approach

As discussed in Chapter 3, the NLSE can be solved analytically in only a few special cases, which means that most analysis consists of numerical simulation (usually by the split step Fourier method; see the end of section 3.4). On the other hand, closed form expressions, such as (3.43), are useful simply because they can be manipulated. A prototype application might be to express some key pulse characteristic such as width in terms of various parameters (dispersion constant, propagation distance, etc.) and then to optimize the former with respect to one or more of the latter. The variational approach to nonlinear pulse propagation, which has been in wide use since its introduction in 1983 [And], provides an approximate, but analytical, expression for the envelope $A(z, t)$ and enables formulaic operations. The purpose of this chapter is to study the variational method and some of its applications. The approach is based on assigning to $A(z, t)$ a trial functional form (or *ansatz*) that is technically correct only in one ideal case but is nevertheless effective in predicting actual pulse behavior over a wide range of pulse shapes.

Much of this chapter is physically motivated and heuristic, but it presents a practical and widely used tool in the study of nonlinear pulse propagation. Chapter 5, which takes up solitons and the inverse scattering method, begins with (3.51) and is independent of this chapter.

The mathematical foundation of the variational approach is covered in section 4.1 and applications to dispersion management are considered in section 4.2. Section 4.3 provides a physical motivation for solitons through the variational method.

The recent article of Malomed [Mal] is a wide-ranging account of the variational approach in nonlinear fibers and related fields. The applications in sections 4.2 and 4.3 are not covered in [Mal].

4.1 Background and Applications

Before discussing specifics, an orientation may be helpful. As Chapter 3 showed, Gaussian input pulses remain Gaussian during propagation in the ideal linear case. Moreover, the Gaussian input shape is a realistic pulse profile and is often assumed in practice [Ag1]. Finally, experiments and numerical simulations show that Gaussian input pulses remain approximately Gaussian during propagation for realistic *nonlinear* fibers under conditions of normal dispersion and low nonlinearity. That is, even though an initially Gaussian pulse spreads as it propagates, the classical bell-shaped curve retains its overall shape but with gradually evolving proportions of height and width. Under these conditions it makes sense as an approximation procedure to postulate that a Gaussian input shape remains Gaussian for all z and to try to write down governing equations for the width, amplitude, and phase. The most effective way to go about this turns out to be through a variational approach, which will be described in this section. Similar remarks apply for the hyperbolic secant pulse profile, but only the Gaussian will be considered here. Gaussian and hyperbolic secant shapes have the very convenient property that their Fourier transforms are also Gaussian and hyperbolic secant, respectively.

4.1.1 Lagrangian and Euler Equations

The calculus of variations associates differential equations with real-valued functionals in such a way that the solutions of the equation coincide with extremals of the functionals. The functional is constructed using a *Lagrangian* corresponding to the equation. Most of the familiar equations of mathematical physics have Lagrangians. For example, the Sturm–Liouville equation $-(py')' + qy = 0$ has the *energy functional*

$$J(y) = \int_a^b (py'^2 + qy^2)dx \tag{4.1}$$

with Lagrangian $L(x, y, y') = p(x)y'^2 + q(x)y^2$. In general, a necessary condition that the function y be a minimum for the functional $J(y) = \int_a^b L(x, y(x), y'(x))dx$ is [Log]

$$\frac{\partial L}{\partial y} - \frac{\partial}{\partial x}\left(\frac{\partial L}{\partial y'}\right) = 0, \tag{4.2}$$

where (4.2) is known as *Euler's equation* and reduces to the original differential equation. For example, with $L = py'^2 + qy^2$, (4.2) reduces to the Sturm–Liouville expression $-(py')' + qy = 0$.

The variational calculus extends to coupled systems of partial differential equations. In two dimensions the functionals take the form

$$J(u, v) = \iint_R L(x, y, u, v, u_x, v_x, u_y, v_y)dxdy,$$

where $u = u(x, y)$, $v = v(x, y)$, and the subscripts denote partial derivatives. A necessary condition that the pair (u, v) be a minimum for $J(u, v)$ is that [Log]

$$\begin{aligned}\frac{\partial L}{\partial u} - \frac{\partial L_{u_x}}{\partial x} - \frac{\partial L_{u_y}}{\partial y} &= 0,\\ \frac{\partial L}{\partial v} - \frac{\partial L_{v_x}}{\partial x} - \frac{\partial L_{v_y}}{\partial y} &= 0,\end{aligned} \tag{4.3}$$

where $L_{u_x} = \partial L/\partial u_x$, etc. The key insight of Anderson [And] was to find a Lagrangian for a system of equations equivalent to the NLSE (3.1). In fact, let

$$L(u, v) = \frac{i}{2}(uv_x - vu_x) - \frac{\beta_2}{2}u_y v_y - \frac{\gamma}{2}u^2v^2 \tag{4.4}$$

and make the identification $x = z, y = t, u = A, v = A^*$. It is straightforward to show that (4.3) is equivalent to (3.1) and its complex conjugate, while (4.4) reduces to

$$L = \frac{i}{2}(AA_z^* - A^*A_z) - \frac{\beta_2}{2}A_tA_t^* - \frac{\gamma}{2}A^2A^{*^2}. \tag{4.5}$$

4.1.2 Gaussian Ansatz and Width Parameter

The variational method is used in fiber optics by combining the Lagrangian with the Ritz procedure [BPJ], in which a particular form, or class of shapes, for the envelope $A(z, t)$ is assumed and the functional $J(A) = \iint_R L(z, t, A, A^*, A_z, A_z^*, A_t, A_t^*)dzdt$ is minimized over the restricted class. As mentioned, the most widely used envelope pattern is the Gaussian. Taking into account (3.39) and (3.41) led Anderson [And] to the Gaussian pulse ansatz

$$A(z, t) = M(z)\exp\left[-t^2\left(\frac{1}{2a(z)^2} - ib(z)\right)\right], \tag{4.6}$$

where $M(z)$, $a(z)$, and $b(z)$ are functions of z only with $M(z)$ complex valued. In (4.6) $M(z)$ represents the pulse center ($t = 0$) complex amplitude, $a(z)$ is the pulse HWHM (section 1.1), and $b(z)$ is the time varying phase or chirp. Inserting (4.6) into (4.5) leads to the realization of $J(A)$ as a real-valued functional that may be viewed, in the calculus of variations sense, as an expression in the independent variables $M(z)$, $M(z)^*$, $a(z)$, and $b(z)$. If the derivatives of $J(A)$ with respect to these four variables are set to 0 and the resulting equations are simplified, they reduce to a set of two coupled equations for the width and chirp, namely [And, BPJ],

$$\begin{aligned} a_z &= -2\beta_2 a(z)b(z), \\ a(z)b_z &= 2\beta_2 a(z)b(z)^2 - \frac{\beta_2}{2a(z)^3} - \frac{\gamma E_0}{2\sqrt{2}a(z)^2}, \end{aligned} \tag{4.7}$$

where E_0 is a constant. The derivation is straightforward but lengthy and will not be repeated here. One fact that emerges from the derivation is that $a(z)|M(z)|^2$ is constant; it is given the symbol E_0, which appears in (4.7), that is, $a(z)|M(z)|^2 = E_0$. Here E_0 is a multiple of the pulse energy, as seen from feeding (4.6) into the energy term from section 3.4:

$$E = \int_{-\infty}^{\infty} |A(z, t)|^2 dt = |M(z)|^2 \int_{-\infty}^{\infty} e^{-t^2/a(z)^2} dt = |M(z)|^2 a(z)\sqrt{\pi}.$$

This expression is constant because attenuation has been ignored; compare the discussion at the beginning of section 3.4. Note also from section 3.4 and (4.6) that $|A(0, 0)|^2 = |M(0)|^2 = P_0$, where P_0 is the peak power. Since $a(z)|M(z)|^2 = E_0$ is constant, then

$$E_0 = a(z)|M(z)|^2 = a(0)P_0 = a_0P_0. \tag{4.8}$$

Equations (4.7) actually reduce to a single differential equation. If the first equation of (4.7) is differentiated to obtain $a_{zz} = -2\beta_2(ab_z + a_zb)$, then $a(z)b_z$ can be replaced by the second equation in (4.7). Making this substitution, noting (4.8), and taking advantage of a fortuitous cancellation leads to the second-order nonlinear ordinary differential equation

$$a_{zz} = \frac{\beta_2^2}{a(z)^3} + \frac{\beta_2 a_0 \gamma P_0}{\sqrt{2}a(z)^2} \tag{4.9}$$

for the pulse width parameter $a(z)$. In regard to initial conditions for (4.9), the initial width $a(0)$ will be specified; that is, $a(0) = a_0$ will be given. Comparing (3.41) and (4.6) one sees that $a_0 = T_0$ and also $b(0) = -C/(2T_0^2) = -C/(2a_0^2)$. The first equation of (4.7) now shows that $a_z(0) = \beta_2 C/a_0$, so the initial conditions to accompany (4.9) are

$$a(0) = a_0, \qquad a_z(0) = \beta_2 C/a_0, \qquad a_0 \text{ and } C \text{ given.} \tag{4.10}$$

Thus the initial pulse width a_0 and chirp C are specified, and this uniquely determines the approximate width $a(z)$ in (4.9) for all z. One can solve (4.9) directly, but the solution is in implicit form and somewhat cumbersome [GrY]. The main advantages of (4.9) are in the speed at which it can be solved numerically, the impressive accuracy with which it predicts pulse width, and the analytical formulas that can be deduced from it. Some examples of such analytical formulas will be given presently.

In (4.9) and (4.10) $a(z)$ typically has units of ps (picoseconds), $b(z)$ has units ps^{-2}, a_z is in ps/Km, β_2 is in ps^2/Km, and C is dimensionless.

The ansatz (4.6) involves three functions ($M(z)$, $a(z)$, and $b(z)$), but (4.9) and (4.10) determine only $a(z)$. Once $a(z)$ is known, however, $b(z)$ follows from the first equation of (4.7) and the modulus $|M(z)|$ is obtained from (4.8). The phase of $M(z)$ can also be determined in a simple way but will not be needed here.

To summarize, the idea is to assume a Gaussian input pulse of the form (3.40) and to posit that the shape remains Gaussian during transmission. This does not really happen in practice but comes close enough that the ansatz (4.6) is a nice predictor of actual pulse behavior.

4.1.3 Comparison with Split Step Data

Figure 4.1 [JHL] shows the normalized root mean square (RMS) width of an initially Gaussian pulse as a function of z as computed by the split step method (end of section 3.4), the variational method (4.9), and two approximate analytical formulas that are based on the split step algorithm [Ag1]. The RMS width is defined (for symmetric pulses) by

$$\sigma^2(z) = \frac{\int_{-\infty}^{\infty} t^2 |A(z,t)|^2 dt}{\int_{-\infty}^{\infty} |A(z,t)|^2 dt}$$

and the normalized RMS is $\sigma_n(z) = \sigma(z)/\sigma(0)$. By (4.6) and the identities

$$\int_{-\infty}^{\infty} e^{-x^2/p^2} dx = p\sqrt{\pi}, \qquad \int_{-\infty}^{\infty} x^2 e^{-x^2/p^2} dx = p^3\sqrt{\pi}/2,$$

one calculates that

$$\sigma^2(z) = a^2(z)/2$$

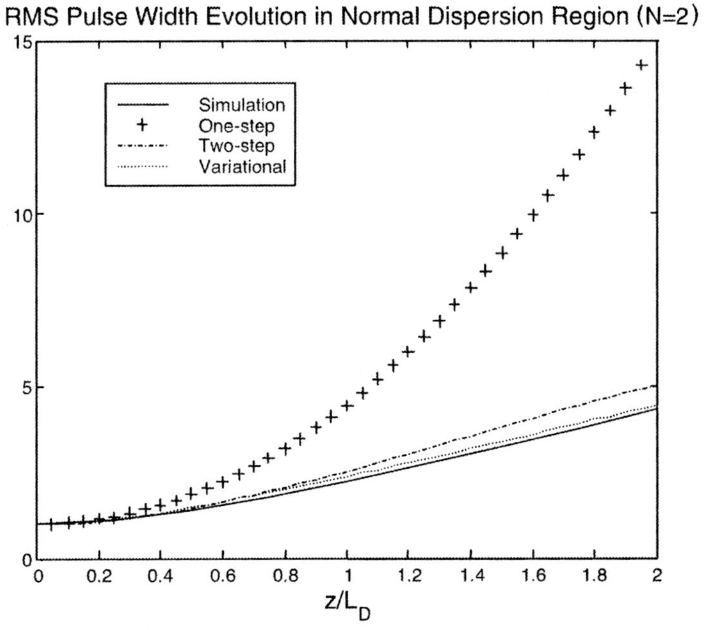

Figure 4.1. *Comparison of normalized RMS pulse width as computed by the split step method (bottom curve, solid), the variational method* (4.9) *(next to bottom curve, dotted), and two approximate analytical methods (dash-dot and "+").*

and thus

$$\sigma_n(z) = \sigma(z)/\sigma(0) = a(z)/a_0.$$

In Figure 4.1 the horizontal scale is normalized length $\zeta = z/L_D$ as defined in section 3.4. Considering the split step data to be exact, note that (4.9) gives an error of only a few percent over two dispersion lengths. The nonlinearity level, as defined in section 3.4, is $N = 2$. In the figure the *one-step method* refers to a single application, over the entire interval, of the split step Fourier approach of applying nonlinearity and dispersion separately and sequentially. The *two-step* method breaks the interval into two halves and applies the split step twice.

In [GrY] (4.9) was used to predict the variation of pulse width and chirp in concatenated lengths of fibers with alternating signs of the dispersion constant β_2. The pulse width for arbitrary z was computed by the split step method and compared to the prediction given by (4.9). The corresponding curves were nearly indistinguishable over the whole range of distances. Extremely close agreement was also found in [BPJ, BJL] and especially [KHEG] in which the propagation distance was 20,000Km (see also [KuE, Geo]).

Note that (4.9) can be solved in closed form if $\gamma = 0$, that is, the ideal linear case, where $a(z)^2$ reduces to T_{out}^2 in (3.43). In other words, the variational approach is exact in the linear case. This observation can be turned into a criterion of sorts: the lower the nonlinearity level, the more accurate the variational method. This means that the normalized nonlinearity parameter N given by $N^2 = a_0^2\gamma P_0/|\beta_2|$ in (3.50) should not be too large if the variational procedure is to be used. A rule of thumb cited in [Ag1] is $N < 1$, but Figure 4.1 (and Figure 4.2 below) show that the variational method yields a good approximation for much larger N. The case $N < 1$, that is, comparatively low energy and short pulses, is of considerable practical interest.

4.1.4 Asymptotically Linear Pulse Broadening

The usefulness of (4.9) can be illustrated with a simple calculation of the asymptotic rate of pulse broadening. For simplicity suppose that $\beta_2 > 0$ and $C \geq 0$ in (4.9) so that $a_{zz} > 0$ in (4.9). Since $a_z(0) \geq 0$, then $a_z(z) > 0$ for all $z > 0$ and, moreover, $a_z(z)$ is an increasing function. In particular, $a(z)$ is unbounded with increasing slope. However, letting $z \to \infty$ in (4.9) implies that $a_{zz}(z) \to 0$ and so, at least intuitively, $a_z(z)$ should approach a constant and therefore the pulse width $a(z)$ would expand linearly with distance eventually. This is consistent with the linear Gaussian model (3.43) but here nonlinearity is permitted as long as N is not too large. In fact, the exact asymptotic growth rate, within the variational picture, can be calculated from (4.9). Multiply both sides of (4.9) by a_z and integrate the result to obtain

$$\frac{1}{2}a_z^2 = -\frac{\beta_2^2}{2a(z)^2} - \frac{\beta_2 a_0\gamma P_0}{\sqrt{2}a(z)} + \frac{1}{2}K,$$

where K is a constant. Simplifying, this becomes

$$a_z^2 = K - \frac{\beta_2^2}{a(z)^2} - \frac{\beta_2 a_0\gamma P_0\sqrt{2}}{a(z)},$$

and the constant K can be evaluated by using (4.10); in fact,

$$a_z^2 = \frac{\beta_2^2(1+C^2)}{a_0^2} + \beta_2\gamma P_0\sqrt{2} - \frac{\beta_2^2}{a(z)^2} - \frac{\beta_2 a_0\gamma P_0\sqrt{2}}{a(z)}. \tag{4.11}$$

The asymptotic linear growth rate of pulse width $a(z)$ is thus given by

$$a_z(\infty)^2 = \frac{\beta_2^2(1+C^2)}{a_0^2} + \beta_2\gamma P_0\sqrt{2}. \tag{4.12}$$

Note that (4.12) reduces to the linear case, that is, the coefficient of z^2 in (3.43), when $\gamma = 0$.

4.1.5 Asymptotic Form of the Spectrum

As a pulse broadens, the power levels decrease. Physically speaking, this implies that the role of nonlinearity should become smaller in relation to the role of dispersion. Recalling that the pulse spectrum in the pure linear case is invariant, this leads to the conjecture that the spectrum should asymptote in the presence of steady pulse broadening. As (4.11) and (4.12) show, this should occur in the normal dispersion case $\beta_2 > 0$. Indeed, there is an asymptotic limit of the spectral width as $z \to \infty$ in normal dispersion. This will now be derived.

The transform of (4.6) can be found using (3.37) and putting

$$-\frac{1}{2\Gamma} = -\frac{1}{2a(z)^2} + ib(z),$$

from which one has

$$\frac{\Gamma}{2} = \frac{a^2}{2(1+4a^2b^2)} + i\frac{a^4b}{1+4a^2b^2}.$$

Using (3.37) the transform $\hat{A}(z,\omega)$ is thus

$$\hat{A}(z,\omega) = M(z)\sqrt{2\pi\Gamma}\exp\left[-\frac{a^2\omega^2}{2(1+4a^2b^2)}\right]\exp\left[-i\frac{a^4b\omega^2}{1+4a^2b^2}\right]. \tag{4.13}$$

It will be convenient to use the normalized RMS [Ag1]

$$\tilde{\sigma}_\omega^2(z) = \frac{\int_{-\infty}^{\infty}\omega^2|\hat{A}(z,\omega)|^2 d\omega}{\int_{-\infty}^{\infty}|\hat{A}(z,\omega)|^2 d\omega} \tag{4.14}$$

to measure the spectral width. Note that (4.14) is a normalized version of (3.3) in which constant factors in $\hat{A}(z,\omega)$ cancel. From (4.13), one has

$$|\hat{A}(z,\omega)|^2 = M(z)\sqrt{2\pi\Gamma}\exp\left[-\frac{a^2\omega^2}{1+4a^4b^2}\right],$$

and, similar to previous calculations, it follows that

$$\tilde{\sigma}_\omega^2(z) = \frac{1}{2}\frac{1+4a^4b^2}{a^2} = \frac{1}{2}\left(4a^2b^2 + \frac{1}{a^2}\right).$$

The term $4a^2b^2$ can be replaced by a_z^2/β_2^2 after squaring the first equation in (4.7). This gives

$$\tilde{\sigma}_\omega^2(z) = \frac{1}{2}\left(\frac{a_z^2}{\beta_2^2} + \frac{1}{a^2}\right),$$

and after substituting from (4.11) there follows the important formula

$$2\tilde{\sigma}_\omega^2(z) = \frac{1+C^2}{a_0^2} + \frac{\sqrt{2}(\gamma P_0)}{\beta_2}\left(1 - \frac{a_0}{a(z)}\right). \tag{4.15}$$

Figure 4.2 [JLS] shows plots of the normalized RMS spectral width as a function of z for N values ranging from $N = 1$ to $N = 5$ and with $C = 0$. Note that the curves appear to asymptote. In the figure the solid curve corresponds to the essentially exact numerical solution and the broken curve to (4.9). Moreover, plots of the spectral amplitude $|\hat{A}(z,\omega)|^2$ (not shown here; see [JLS]) are also in close agreement.

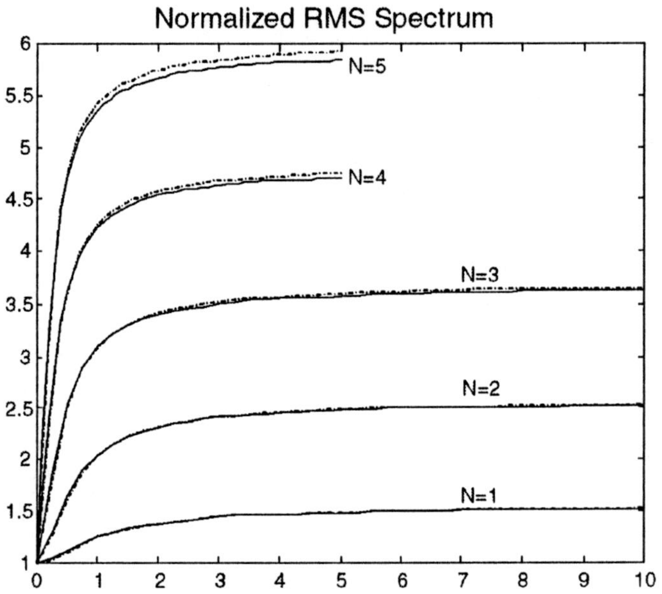

Figure 4.2. *Normalized RMS spectral width by split step (solid line) and* (4.9) *(broken line). Horizontal axis is the same variable as in Figure* 4.1.

The above derivations were carried out under the assumption $\beta_2 > 0$. Versions of (4.12) and (4.15) are also valid for the anomalous dispersion case $\beta_2 < 0$, but there is a caveat that the normalized nonlinearity constant N cannot exceed an explicit constant, the value of which will be discussed later. The situation is that when $\beta_2 < 0$ soliton solutions of (3.48) can form, in which case the pulse asymptotically evolves into a precisely known form that is different from (4.6) or its hyperbolic secant analog. This will be covered in section 4.3.

4.2 Dispersion Management

The purpose of this section is to discuss ways in which the sign of the constant β_2 and pulse chirp can be used to compensate for fiber dispersion. The freedom to choose the sign of the dispersion constant β_2 allows one to reverse the pulse broadening over a span of fiber by splicing into a second fiber with the opposite sign of β_2. Under appropriate conditions it is possible for the output width of the second fiber to equal the input width of the first fiber. In this situation one says that the second fiber *compensates* for the dispersion of the first fiber. Also, by using an appropriate chirp C in (4.10), an input pulse can be made to initially compress, and then expand, so that the output and input widths in a single span of fiber are equal. This process is sometimes called *precompensation*. Techniques such as these comprise the field of *dispersion management*.

4.2.1 Dispersion Maps

First consider the idea of concatenating sections of fiber with opposite signs of dispersion. To explain the concept, go back to the ideal linear case $\gamma = 0$ and write down the solution (3.2) in the Fourier domain. Imagine a segment of fiber with dispersion constant $\beta_2^{(1)}$ and length L_1. The transform of the output envelope is then

$$\hat{A}(L_1, \omega) = \hat{A}(0, \omega)e^{i\beta_2^{(1)}\omega^2 L_1/2}.$$

If the pulse is now fed into a second fiber of length L_2 with dispersion constant $\beta_2^{(2)}$, then the output at length z in the second segment is

$$\hat{A}(z, \omega) = \hat{A}(0, \omega)e^{i\beta_2^{(1)}\omega^2 L_1/2}e^{i\beta_2^{(2)}\omega^2(z-L_1)/2}. \tag{4.16}$$

Putting $z = L_1 + L_2$ in (4.16) gives the output of the second link as

$$\hat{A}(L_1 + L_2, \omega) = \hat{A}(0, \omega)e^{i\omega^2(\beta_2^{(1)}L_1+\beta_2^{(2)}L_2)/2}.$$

The input and output pulses are therefore identical if

$$\beta_2^{(1)}L_1 + \beta_2^{(2)}L_2 = 0. \tag{4.17}$$

If the fiber links have oppositely signed dispersion constants and (4.17) can be arranged, then the second link perfectly compensates for dispersion in the first link. The configuration of constants in (4.17) is called a *dispersion map*.

Of course, (4.17) comes from assuming $\gamma = 0$, but the variational method provides a way of extending the dispersion map to more general fibers. The principle tool is a series expansion of the width parameter $a(z)^2$ in (4.6). This is derived next.

4.2.2 Dispersion Maps by the Variational Method

The asymptotic limit (4.12) suggests that $a(z)^2$ should have order of magnitude z^2 for large z. Moreover, (4.9) provides a means of calculating higher order derivatives of $a(z)$. With this in mind expand the squared width into the power series

$$a(z)^2 = c_0 + c_1 z + c_2 z^2 + c_3 z^3 + c_4 z^4 + \cdots \tag{4.18}$$

and evaluate the coefficients by successive application of (4.9). Setting $z = 0$ in (4.18) gives $c_0 = a_0^2$. Differentiating (4.18) to obtain

$$2aa_z = c_1 + 2c_2 z + 3c_3 z^2 + \cdots, \tag{4.19}$$

setting $z = 0$, and using (4.10) then results in

$$c_1 = 2a(0)a_z(0) = 2a_0\left(\frac{\beta_2 C}{a_0}\right) = 2\beta_2 C.$$

The derivative of (4.19) is

$$2aa_{zz} + 2a_z^2 = 2c_2 + 6c_3 z + \cdots, \tag{4.20}$$

in which a_{zz} can be replaced by (4.9). Carrying this out, setting $z = 0$, using (4.10), and simplifying gives the quadratic coefficient as

$$c_2 = \frac{\beta_2^2(1+C^2)}{a_0^2} + \frac{\beta_2 \gamma P_0}{\sqrt{2}}.$$

Higher order coefficients may be computed by successive iteratives of (4.9); the series (4.18) through two terms is [BPJ]

$$a(z)^2 \cong a_0^2 + 2\beta_2 C z + \left[\frac{\beta_2^2(1+C^2)}{a_0^2} + \frac{\beta_2 \gamma P_0}{\sqrt{2}}\right] z^2. \tag{4.21}$$

Note that (4.21) reduces to (3.43) in the ideal linear case $\gamma = 0$.

The series (4.21) is valid over an interval $0 < z < L_1$, where $\beta_2 = \beta_2^{(1)}$. The idea behind using the variational method to calculate dispersion maps is to generate the analog of (4.21) on an interval $L_1 < z < L_1 + L_2$, where $\beta_2 = \beta_2^{(2)}$, with $\beta_2^{(1)}$ and $\beta_2^{(2)}$ having opposite signs, and to match the output at $z = L_1 + L_2$ to that at $z = 0$. This corresponds physically to two concatenated fibers, with lengths L_1 and L_2 and opposite signs of dispersion, respectively, extending over $0 < z < L_1$ and $L_1 < z < L_1 + L_2$. To this end, expand the pulse width in a series

$$a(z)^2 = d_0 + d_1(z - L_1) + d_2(z - L_1)^2 + \cdots \tag{4.22}$$

in the interval $L_1 < z < L_1 + L_2$ and evaluate the coefficients as before. In the coupled equations (4.7), continuity at $z = L_1$ of pulse width and chirp, $a(z)$ and $b(z)$, are physical necessities. But this means in view of (4.7) that the derivative $a_z(z)$ is discontinuous at $z = L_1$ because β_2 changes sign there. Enforcing continuity of $a(z)$ and $b(z)$ in (4.7) implies that the jump discontinuity of $a_z(z)$ is given by

$$\frac{a_z(L_1+)}{a_z(L_1-)} = \frac{\beta_2^{(2)}}{\beta_2^{(1)}}, \tag{4.23}$$

where $L_1\pm$ refers to the right (upper sign) and left (lower sign) limits at L_1. Equation (4.23) shows, for example, that if the pulse width is increasing up to $z = L_1$ but the pulse is fed into a fiber with the opposite sign of dispersion, then the width immediately begins to decrease. That is, reverse dispersion begins.

To calculate the coefficients in (4.22), note first that setting $z = L_1$ in (4.22) gives $d_0 = a(L_1)^2 = a_1^2$. Then taking the derivative, evaluated at $z = L_1$, gives

$$d_1 = 2aa_z(L_1+) = 2aa_z(L_1-)\frac{\beta_2^{(2)}}{\beta_2^{(1)}} \tag{4.24}$$

by (4.23). In (4.24) the term $2aa_z(L_1-)$ is evaluated by (4.21) and is given by

$$2aa_z(L_1-) \cong 2\beta_2 C + \left[\frac{\beta_2^2(1+C^2)}{a_0^2} + \frac{\beta_2 \gamma P_0}{\sqrt{2}}\right](2L_1).$$

Substituting this into (4.24) and simplifying yields

$$d_1 = 2\beta_2^{(2)}C + 2L_1\left(\frac{\beta_2^{(1)}\beta_2^{(2)}(1+C^2)}{a_0^2} + \frac{\beta_2^{(2)}\gamma P_0}{\sqrt{2}}\right). \tag{4.25}$$

The quadratic coefficient d_2 is rather complicated but can be put into fairly simple form if nonlinearity is neglected in the second fiber. This is a reasonable assumption because peak power will have been reduced by pulse broadening in the first fiber. Dispersion can therefore be assumed to dominate nonlinearity in the second fiber. Proceeding in this way one calculates the series to quadratic terms as

$$a(z)^2 = a_1^2 + d_1(z-L_1) + \left(\frac{(\beta_2^{(2)})^2}{a_1^2} + \frac{d_1^2}{4a_1^2}\right)(z-L_1)^2 + \cdots, \qquad z > L_1, \tag{4.26}$$

with d_1 given by (4.25) and $a_1 = a(L_1)$.

To illustrate (4.26), suppose for simplicity that $\beta_1^{(1)} > 0$ and $C = 0$. Then d_1 in (4.25) reduces to $d_1 = 2L_1\beta_2^{(2)}((\beta_2^{(1)}/a_0^2) + (\gamma P_0/\sqrt{2}))$, which is negative if $\beta_2^{(2)} < 0$. This means that the width $a(z)$ in (4.21) increases up to $z = L_1$ in the first fiber but then begins to decrease as it passes into the second fiber in (4.26). Note that the decrease is temporary because the quadratic coefficient in (4.26) is positive. However, this calculation shows that the principle of dispersion compensation as expressed in (4.17) for the ideal linear fiber should carry over to the nonlinear case.

4.2.3 The Minus One-Half Rule

It is very interesting to find the analog of (4.17), within the variational approximation, for nonlinear propagation. To this end consider two concatenated fibers of lengths L_1 and L_2, dispersion constants $\beta_1^{(1)} > 0$ and $\beta_1^{(2)} < 0$, and let a_1 and a_2 be the output widths of the first and second fibers, respectively. The condition for perfect dispersion compensation is $a_2 = a_0$. From (4.21) and (4.26),

$$\begin{aligned} a_1^2 &\cong a_0^2 + 2\beta_2^{(1)}CL_1 + \left[\frac{(\beta_2^{(1)})^2(1+C^2)}{a_0^2} + \frac{\beta_2^{(1)}\gamma P_0}{\sqrt{2}}\right]L_1^2, \\ a_2^2 &\cong a_1^2 + d_1L_2 + \left(\frac{(\beta_2^{(2)})^2}{a_1^2} + \frac{d_1^2}{4a_1^2}\right)L_2^2, \end{aligned} \tag{4.27}$$

where nonlinearity has again been neglected in the second fiber. Define

$$\Delta_1 = \beta_2^{(1)}L_1,\ \Delta_2 = \beta_2^{(2)}L_2,\ F_1 = \gamma P_0 L_1/\sqrt{2}$$

so that the first equation of (4.27) is

$$\begin{aligned} a_1^2 &\cong a_0^2 + 2\Delta_1 C + \frac{\Delta_1^2(1+C^2)}{a_0^2} + \frac{\Delta_1 L_1 \gamma P_0}{\sqrt{2}} \\ &= a_0^2 + \frac{\Delta_1^2(1+C^2)}{a_0^2} + \Delta_1(2C + F_1) \end{aligned} \tag{4.28}$$

and the second is

$$a_2^2 \cong a_1^2 + \frac{\Delta_2^2 + (d_1 L_2/2)^2}{a_1^2} + d_1 L_2, \tag{4.29}$$

where by (4.25)

$$\begin{aligned} d_1 L_2 &= 2\Delta_2 C + \frac{2\Delta_1\Delta_2(1+C^2)}{a_0^2} + \frac{2L_1\Delta_2\gamma P_0}{\sqrt{2}} \\ &= \frac{2\Delta_1\Delta_2(1+C^2)}{a_0^2} + 2\Delta_2(C+F_1). \end{aligned} \tag{4.30}$$

Multiplying (4.28) by a_0^2 and (4.29) by a_1^2 to clear the denominators, one obtains

$$\begin{aligned} a_1^2 a_0^2 &\cong a_0^4 + \Delta_1^2(1+C^2) + a_0^2\Delta_1(2C+F_1), \\ a_1^2 a_2^2 &\cong a_1^4 + \Delta_2^2 + (d_1 L_2/2)^2 + a_1^2 d_1 L_2. \end{aligned} \tag{4.31}$$

For perfect dispersion compensation (i.e., $a_2 = a_0$) the expressions in (4.31) are equal. Equating in (4.31), using (4.30) to substitute for $d_1 L_2$, and replacing a_1^2 by (4.28) yields

$$\begin{aligned} &a_0^4 + \Delta_1^2(1+C^2) + a_0^2\Delta_1(2C+F_1) \\ &= \left(a_0^2 + \frac{\Delta_1^2(1+C^2)}{a_0^2} + \Delta_1(2C+F_1)\right)^2 + \Delta_2^2 + (d_1 L_2/2)^2 \\ &\quad + \left(a_0^2 + \frac{\Delta_1^2(1+C^2)}{a_0^2} + \Delta_1(2C+F_1)\right)\left(\frac{2\Delta_1\Delta_2(1+C^2)}{a_0^2} + 2\Delta_2(C+F_1)\right). \end{aligned}$$

Performing some simplifications, this reduces to

$$\begin{aligned} \Delta_1 a_0^2\left(\frac{\Delta_1(1+C^2)}{a_0^2} + (2C+F_1)\right) &= \Delta_2^2 + 2a_0^2 X + X^2, \\ X = (\Delta_1+\Delta_2)\left(\frac{\Delta_1(1+C^2)}{a_0^2} + (C+F_1)\right) &+ C\Delta_1. \end{aligned} \tag{4.32}$$

In (4.32) one can solve for $\Delta_2 = \beta_2^{(2)} L_2$ in terms of $\Delta_1 = \beta_2^{(1)} L_1$ and the other parameters in order to predict the appropriate dispersion and distance in the second fiber needed to compensate for dispersion in the first fiber.

There turns out to be a simple way to test (4.32), namely, its asymptotic form in the limit of large input width a_0. This is because (4.32) simplifies dramatically if one lets $a_0 \to \infty$ and retains only the dominant terms of (4.32). It is easy to see that the dominant terms are

$$\Delta_1(2C+F_1) = 2[(\Delta_1+\Delta_2)(C+F_1) + C\Delta_1],$$

from which Δ_2 can be found as

$$\Delta_2 = -\frac{\Delta_1(F_1+2C)}{2(C+F_1)}. \tag{4.33}$$

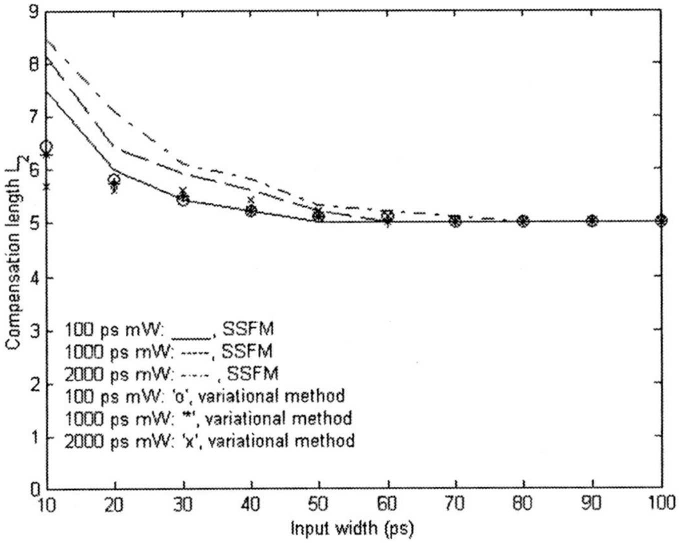

Figure 4.3. *Compensation length L_2 plotted against input width.*

The unchirped case $C = 0$ is especially interesting since (4.33) reduces to the *minus one-half rule*

$$\Delta_2 = -\Delta_1/2, \tag{4.34}$$

which is interesting to compare with (4.17), namely, $\Delta_2 = -\Delta_1$, in the ideal linear case.

Although (4.34) is an asymptotic result, it actually is rather accurate over a wide range of pulse parameters. Figure 4.3 [JSW1] plots the compensation lengths L_2 against pulse widths with $C = 0$ when $L_1 = 100Km$, $\beta_2^{(1)} = 2ps^2/Km$, and $\beta_2^{(2)} = -20ps^2/Km$, and for a range of energy levels. Since $\Delta_1 = 200ps^2$ and $\beta_2^{(2)} = -20ps^2/Km$, then (4.34) predicts $L_2 \cong 5Km$ for large a_0; the asymptotic tendency towards $5Km$ is clearly indicated in the figure. The marks (o, *, x) correspond to the calculation of compensation length by (3.32) and the curves (solid, dash, dash-dot) represent calculation by the split step method. The total energy as given by (4.8) is held fixed for each curve in order to provide a benchmark. That is, changing a_0 while holding peak power P_0 fixed changes the pulse energy by (4.8). Note that the compensation length is relatively insensitive to total energy, as expected from the cancellation of the F_1 term in (4.33) when $C = 0$. Energy levels of $100psmW$ to $2000psmW$ are significant, that is, not close to the linear case where (4.17) holds. In terms of the physical interpretation of (4.34), nonlinear effects dominate in the first half of the first fiber; the spectrum expands, but the pulse broadening is minimal. The increase in spectral width causes dispersion to dominate in the second half of the first fiber. Therefore, roughly speaking, the second fiber has to compensate for dispersion in only half of the first fiber.

4.2.4 Precompensation and Optimum Chirp

The truncated series (4.21) provides a simple way of understanding another aspect of dispersion management. As already discussed, and as (4.21) shows, the width $a(z)$ eventually increases linearly under normal dispersion $\beta_2 > 0$. However, note that if $C < 0$, the derivative of (4.21) is negative for small z. This says that under normal dispersion a negatively chirped pulse initially compresses; that is, for some propagation distance the pulse width will be less than the initial width. Call this distance the *compression length*, L_{com}. Then for $0 < z < L_{com}$ one knows that the pulse does not broaden beyond its own timing window. This is a form of precompensation by prechirping and is induced by the condition $C < 0$.

How large is L_{com}, and can it be specified in advance? To answer this all one has to do is set $a(z) = a_0$ in (4.21) and solve the resulting analytical expression for the key parameters. It will facilitate matters to use normalized variables and, to this end, let $\zeta = z/L_D$ ($L_D = a_0^2/|\beta_2|$) as in section 3.4, where for simplicity $\beta_2 > 0$ will again be assumed. Dividing (4.21) by a_0^2 gives

$$\frac{a(z)^2}{a_0^2} \cong 1 + 2C\zeta + (1 + C^2)\zeta^2 + \frac{\beta_2 z^2 \gamma P_0}{\sqrt{2}a_0^2}. \tag{4.35}$$

From section 3.4 the last term is

$$(\beta_2 z^2 \gamma P_0 / a_0^2\sqrt{2}) = (\zeta^2 L_D^2 / L_D L_{NL}\sqrt{2}) = \zeta^2 N^2/\sqrt{2}.$$

Setting $a(z) = a_0$ and dividing by ζ^2 in (4.35) yields a quadratic equation in the chirp parameter C,

$$C^2 + \frac{2C}{\zeta} + \left[1 + \frac{N^2}{\sqrt{2}}\right] = 0. \tag{4.36}$$

Solving the quadratic equation (4.36) for C then gives

$$C = \frac{-1 + \sqrt{1 - \zeta^2\left(1 + (N^2/\sqrt{2})\right)}}{\zeta}, \tag{4.37}$$

where the minus sign has been selected in the quadratic formula to minimize $|C|$. A small value of $|C|$ is desirable in order not to broaden the spectrum any more than necessary; see (4.15). Equation (4.37) gives a real value of C provided that

$$\zeta^2\left(1 + \frac{N^2}{\sqrt{2}}\right) \leq 1. \tag{4.38}$$

In terms of physical variables (4.38) is

$$z \leq \frac{a_0^2}{\beta_2\left(1 + \frac{N^2}{\sqrt{2}}\right)\sqrt{2}}. \tag{4.39}$$

For any set of parameters satisfying (4.39) the input and output pulse widths can be made equal if the pulse is chirped according to (4.37). The maximum normalized compression distance is thus

$$\zeta_{\max} = \left(1 + \frac{N^2}{\sqrt{2}}\right)^{-1} \tag{4.40}$$

and the corresponding optimal C is $C = -1/\zeta_{\max}$. Note that $\zeta_{\max} = 1$ and $C = -1$ in the ideal linear case $N = 0$.

Figure 4.4 shows the optimal C values plotted against normalized distance ζ for selected values of N. The solid lines represent the chirp C as calculated by trial and error using the split step method; the symbols "o" correspond to the values given by (4.37). The values obtained from (4.37) were used to begin the split step search for the optimum. Note that there is nothing in the governing equation (3.1) to indicate that an optimum chirp even exists, much less to predict its value. In the trial and error split step procedure, for a given distance and initial width, a trial chirp value is selected and a simulation is run to find the output pulse width. The values of C are adjusted until the output and input widths are approximately the same. This is a lengthy and inefficient procedure made possible only by a judicious initial estimate of C provided by (4.37). As Figure 4.4 indicates, the initial guess is actually very close to the exact chirp required [JaS].

To put the discussion in context, chirping can be viewed as one means of controlling dispersion. In the above discussion, setting the input and output widths equal is an arbitrary but natural benchmark. It should be mentioned that there are disadvantages to pulse chirping, for example, the expanded spectrum as seen in (4.15). In this connection, note for $\beta_2 > 0$ the curious phenomenon that the pulse width $a(z)$ and spectral width $\tilde{\sigma}_\omega(z)$ increase or decrease together by (4.15). Finally, the above analysis partially carries over to the anomalous dispersion case $\beta_2 < 0$ but is more sensitive to the size of the nonlinearity parameter N [JaS].

4.2.5 Optimum Input Width

The section now closes with a brief discussion of optimal input width. The context here is to expand upon the discussion below (3.43), where an input width $T_0 = a_0$ was chosen to minimize the corresponding output width T_{out} in a linear fiber. The optimum input width was found to be $a_{0.crit}^2 = \beta_2 z\sqrt{1 + C^2}$. How does this change when nonlinearity is taken into account? The tool seems at first to be in hand in the form of (4.21), which can be minimized with respect to the input a_0. Note, however, that the nonlinear term γP_0 drops out when differentiating with respect to a_0. That is, the optimum using (4.21) is the same as that calculated in (3.43) and does not depend on nonlinearity.

One way around this is to include more terms in (4.18). A calculation shows that (4.18) to four terms is given by

$$\begin{aligned} a(z)^2 &= a_0^2 + 2\beta_2 C z + \left[\frac{\beta_2^2(1 + C^2)}{a_0^2} + \frac{\beta_2 \gamma P_0}{\sqrt{2}}\right] z^2 \\ &\quad + \frac{\beta_2^2 \gamma P_0 C}{3\sqrt{2} a_0^2} z^3 + \left[\frac{\beta_2 \gamma P_0}{12\sqrt{2}} \left(\frac{\beta_2^2}{a_0^4} + \frac{\beta_2 \gamma P_0}{\sqrt{2} a_0^2} - \frac{2\beta_2^2 C^2}{a_0^4}\right)\right] z^4 + \cdots, \end{aligned} \tag{4.41}$$

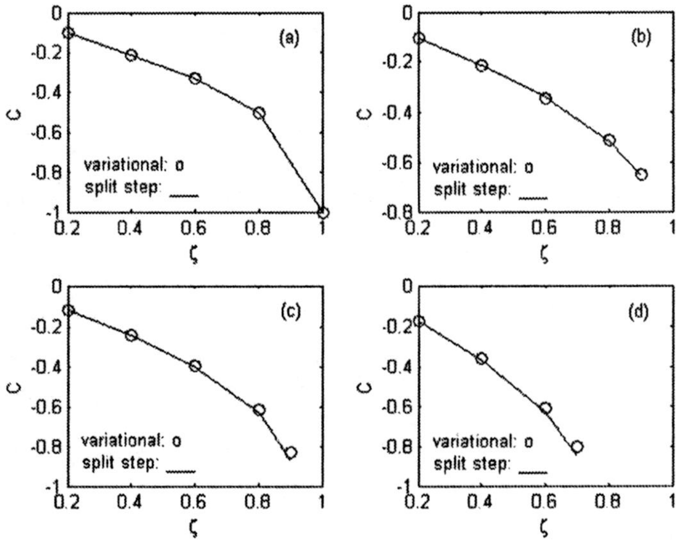

Figure 4.4. *Comparison of the variational ("o") and split step (solid lines) calculations of the value of C required to insure equal input and output widths at the dimensionless propagation distance ζ. Parameters are* (a) $N = 0$, $\zeta_{\max} = 1$; (b) $N = .2$, $\zeta_{\max} = .98$; (c) $N = .5$, $\zeta_{\max} = .92$; *and* (d) $N = 1$, $\zeta_{\max} = .77$.

where nonlinear terms are now involved after differentiating with respect to a_0. The optimum input is the solution of a cubic equation all of whose roots can be found using symbolic software [BPJ]. Note that the asymptotic form of $a(z)$ in (4.41) for large z is incorrect; (4.21) shows that the squared width is asymptotically quadratic rather that quartic. However, the optimum is calculated for a fixed distance and [BPJ] shows that (4.41) provides an accurate prediction of the optimum.

An interesting sidelight is that (4.41) could be extended to include sixth-order terms, but the resulting equation for the optimum is then a quintic polynomial that cannot be factored in general [BPJ].

4.2.6 Variational Method with Loss Term

Most of the results of this section can be extended so as to include effects of fiber loss. This proceeds by using the idea of average power, as mentioned below (3.50) in Chapter 3. Suppose the loss term corresponding to the attenuation coefficient α from (3.46) is included in (3.1). Let P_0 be the peak power as before and define the average power by

$$P_{avg} = (1/z)\int_0^z P_0 e^{-\alpha s} ds = P_0(e^{-\alpha z} - 1)/\alpha z.$$

Let $\tilde{L}_{NL} = 1/(\gamma P_{avg})$ be the modified nonlinear length using P_{avg} instead of P_0, and define $\tilde{N}^2 = L_D/\tilde{L}_{NL}$ as an alternative to the normalized nonlinear index N. Considering, for simplicity, the normal dispersion case $\beta_2 > 0$, the quadratic series (4.21) becomes [JSW2]

$$\frac{a^2(z)}{a_0^2} \cong 1 + 2C\left(\frac{z}{L_D}\right) + \left(1 + C^2 + \frac{\tilde{N}^2}{\sqrt{2}}\right)\left(\frac{z}{L_D}\right)^2.$$

If this series is used in the calculation of (4.37), then the optimal chirp can be predicted accurately, the basis being the variational formula (4.35). In fact, the above series gives a very good approximation to the actual pulse width when loss is taken into account [JSW2]. There is also a variational ansatz for the NLSE with both diffusion and attenuation terms [KaM].

4.3 Optical Solitons in the Variational Context

This section studies solitons within the variational framework. Optical solitons are theoretical solutions of the focusing case NLSE in (3.51) that are either periodic in the distance variable or, in the case of the fundamental soliton, maintain their shape exactly, without pulse or spectral broadening, in an ideal lossless fiber. Within the variational context it will be shown that *fundamental* solitons represent a perfect balance of the dispersion and nonlinear terms in (3.1) whereby the effects of dispersion and nonlinearity cancel, rather than combine, and thereby prevent pulse broadening.

Since fibers have attenuation, solitons do not really exist. However, the *soliton effect*, the tendency to balance but without complete balancing, is a physical phenomenon that occurs in practice and forms the basis of *dispersion-managed soliton systems*. Solitons were first predicted in 1973 in a paper of Hasegawa and Tappert; the effect was experimentally observed in 1980 at Bell Laboratories by Mollenauer, Stolen, and Gordon [Ag1].

Optical solitons will be covered in detail in Chapter 5. The overarching fact is that in the anomalous dispersion (focusing) case all solutions of (3.51) asymptotically approach exactly computable functional forms on intervals $[a, b]$ as $\zeta \to \infty$. The asymptotic limits are combinations of hyperbolic secants and are completely determined by the initial data $u(0, t)$. If the initial data itself is a soliton, then the pulse propagates without change in shape or spectrum.

However, it is not clear from (3.51) how the dispersion and nonlinear terms can combine to suppress pulse spreading. Insight into the process is available by the variational method which, as always in nonlinear fiber optics, provides physical motivation but is only approximate. Syntheses will now be given for the properties that solitons can occur only in the anomalous dispersion regime $\beta_2 < 0$ and only if the nonlinearity parameter N in (3.50) is sufficiently large.

4.3.1 Governing Equations in Normalized Form

Expressing the key variational equations of section 4.1 in normalized dimensionless form will simplify steps. The dispersion length $L_D = a_0^2/|\beta_2|$, nonlinear length $L_{NL} = (\gamma P_0)^{-1}$, and nonlinearity constant N given by $N^2 = L_D/L_{NL}$ were introduced in section 3.4. Now define a new dimensionless length variable ξ and normalized pulse width $s = s(\xi)$, respectively, by $\xi = z/L_{NL}$ and

$$s(\xi) = \frac{a(z)}{\sqrt{|\beta_2|L_{NL}}}. \tag{4.42}$$

Putting $\xi = 0$ in (4.42) gives $s_0 = s(0) = a_0/\sqrt{|\beta_2|L_{NL}}$ and by the definition of L_D,

$$s_0^2 = \frac{a_0^2}{|\beta_2|L_{NL}} = \frac{L_D}{L_{NL}} = N^2;$$

that is, $s_0 = N$, the normalized nonlinearity index defined in (3.50). The change of variable (4.42) thus folds the nonlinearity level N into the initial width. To express (4.9) in terms of $s(\xi)$, start by taking the first derivative,

$$s_\xi = \frac{a_z}{\sqrt{|\beta_2|L_{NL}}} z_\xi = \frac{L_{NL}a_z}{\sqrt{|\beta_2|L_{NL}}}.$$

The second derivative is

$$s_{\xi\xi} = \frac{L_{NL}a_{zz}}{\sqrt{|\beta_2|L_{NL}}} z_\xi = \frac{L_{NL}^2 a_{zz}}{\sqrt{|\beta_2|L_{NL}}},$$

and by (4.9)

$$s_{\xi\xi} = \frac{L_{NL}^2}{\sqrt{|\beta_2|L_{NL}}}\left(\frac{\beta_2^2}{a(z)^3} + \frac{\beta_2 a_0 \gamma P_0}{\sqrt{2}a(z)^2}\right). \tag{4.43}$$

Substituting $a(z) = s(\xi)\sqrt{|\beta_2|L_{NL}}$ reduces the first term on the right of (4.43) to

$$\frac{L_{NL}^2}{\sqrt{|\beta_2|L_{NL}}} \frac{\beta_2^2}{(s(\xi)\sqrt{|\beta_2|L_{NL}})^3} = \frac{1}{s(\xi)^3}.$$

As for the second term in (4.43), similar operations yield $s_0\beta_2/(\sqrt{2}|\beta_2|s)$ and so (4.43) becomes

$$s_{\xi\xi} = \frac{1}{s^3} + \frac{\sigma s_0}{\sqrt{2}s^2}, \qquad \sigma = sign(\beta_2), \tag{4.44}$$

where σ, the algebraic sign of β_2, is 1 if $\beta_2 > 0$ and -1 if $\beta_2 < 0$. By an analogous calculation $s_\xi(0)$, which will be important in the analysis, satisfies

$$s_\xi(0) = \frac{\sigma C}{s_0}. \tag{4.45}$$

Finally, the normalized version of (4.11) can be derived as

$$s_\xi^2 = \frac{C^2}{s_0^2} + \left(\frac{1}{s_0^2} - \frac{1}{s^2}\right) + \sigma\sqrt{2}\left(1 - \frac{s_0}{s}\right). \tag{4.46}$$

Having (4.43)–(4.46) in hand, some simple analyses of both the normal and anomalous dispersion cases can now be given.

4.3.2 Normal Dispersion

Here $\beta_2 > 0$ and the key equations are

$$s_{\xi\xi} = \frac{1}{s^3} + \frac{s_0}{\sqrt{2}s^2}, \qquad s_\xi(0) = \frac{C}{s_0}, \qquad s_\xi^2 = \frac{C^2}{s_0^2} + \left(\frac{1}{s_0^2} - \frac{1}{s^2}\right) + \sqrt{2}\left(1 - \frac{s_0}{s}\right). \tag{4.47}$$

Note that the graph of $s(\xi)$ is convex upward since $s_{\xi\xi} > 0$. Consider subcases determined by the sign of C.

(i) $C \geq 0$: From the signs in (4.47), $s(\xi)$ is concave up, increasing, and unbounded. Letting $\xi \to \infty$ in (4.47) yields an asymptotic slope given by $s_\xi^2(\infty) = ((1+C^2)/s_0^2) + \sqrt{2}$.

(ii) $C < 0$: Here $s(\xi)$ is initially decreasing and falls to a local minimum where $s_\xi = 0$. To describe the critical point, write the term s_ξ^2 in (4.47) in the form

$$s_\xi^2 = \frac{1 + C^2 + \sqrt{2}s_0^2}{s_0^2 s^2}\left(s^2 - \frac{\sqrt{2}s_0^3 s}{1 + C^2 + \sqrt{2}s_0^2} - \frac{s_0^2}{1 + C^2 + \sqrt{2}s_0^2}\right),$$

which can be factored as

$$s_\xi^2 = \frac{1 + C^2 + \sqrt{2}s_0^2}{s_0^2 s^2}(s - s_1)(s - s_2),$$

where the roots $s_1 = s_1(C)$ and $s_2 = s_2(C)$ are given by

$$\begin{aligned} s_1 &= \frac{s_0}{\sqrt{2}} \frac{s_0^2 + \sqrt{(s_0^2 + \sqrt{2})^2 + 2C^2}}{1 + C^2 + \sqrt{2}s_0^2} > 0, \\ s_2 &= \frac{s_0}{\sqrt{2}} \frac{s_0^2 - \sqrt{(s_0^2 + \sqrt{2})^2 + 2C^2}}{1 + C^2 + \sqrt{2}s_0^2} < 0. \end{aligned} \tag{4.48}$$

Note for $C = 0$ that $s_1(0) = s_0$. Moreover, it can be shown that $s_1(C)$ is decreasing in C. Thus the critical points in (4.48) satisfy

$$s_2(C) < 0 < s_1(C) < s_0.$$

This means that the graph of $s(\xi)$ starts at s_0, decreases to a minimum $s_1(C)$ at some point ξ_0, and is increasing thereafter with the asymptotic slope $s_\xi(\infty)$ obtained earlier. There is a point ξ_1 where $s(\xi)$ returns to its initial value s_0; this point corresponds to the compression length L_{com} discussed in the previous section in regard to precompensation. The behavior of (4.47) is illustrated in Figure 4.5 with $s_0 = 1$ and plots for $C = 0$ and $C = \pm 1$.

4.3.3 Anomalous Dispersion

Now suppose $\beta_2 < 0$ so that the analog of (4.47) is

$$s_{\xi\xi} = \frac{1}{s^3} - \frac{s_0}{\sqrt{2}s^2}, \qquad s_\xi(0) = -\frac{C}{s_0}, \qquad s_\xi^2 = \frac{C^2}{s_0^2} + \left(\frac{1}{s_0^2} - \frac{1}{s^2}\right) - \sqrt{2}\left(1 - \frac{s_0}{s}\right), \tag{4.49}$$

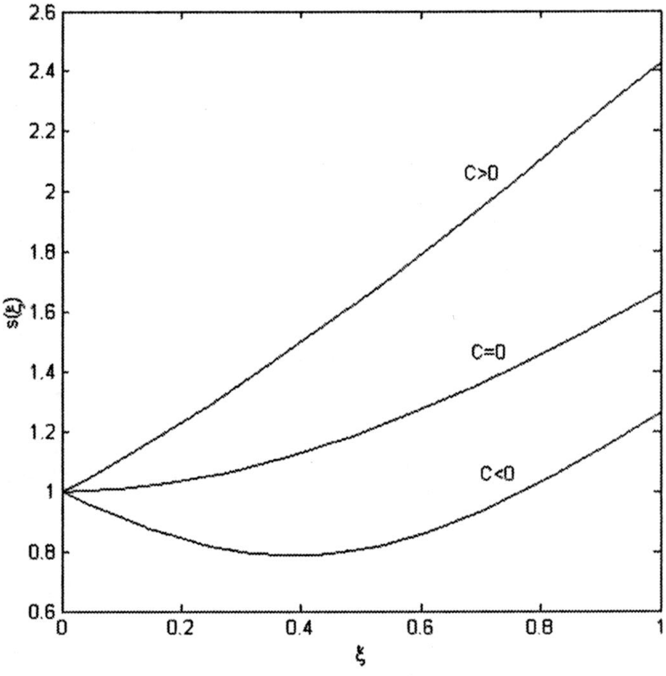

Figure 4.5. *Representative plots of $s(\xi)$ in the normal dispersion case with $s_0 = 1$.*

having again used (4.43)–(4.46). The first of (4.49), which may be written as

$$s_{\xi\xi} = \frac{1}{s^3}\left(1 - \frac{s s_0}{\sqrt{2}}\right),$$

shows that the graph of $s(\xi)$ is convex upward, $s_{\xi\xi} > 0$, if and only if $s < \sqrt{2}/s_0$. The last equation in (4.49) can be factored as before, this time giving

$$s_\xi^2 = \frac{1 + C^2 - \sqrt{2}s_0^2}{s_0^2 s^2}(s - s_1)(s - s_2),$$

where the roots $s_1 = s_1(C)$ and $s_2 = s_2(C)$ are

$$\begin{aligned} s_1 &= \frac{s_0}{\sqrt{2}} \frac{\sqrt{(s_0^2 - \sqrt{2})^2 + 2C^2} - s_0^2}{1 + C^2 - \sqrt{2}s_0^2}, \\ s_2 &= -\frac{s_0}{\sqrt{2}} \frac{\sqrt{(s_0^2 - \sqrt{2})^2 + 2C^2} + s_0^2}{1 + C^2 - \sqrt{2}s_0^2}. \end{aligned} \tag{4.50}$$

These equations suggest proceeding by cases, with a first case being $s_0^2 < 1/\sqrt{2}$, so that the denominators in (4.50) satisfy $1 + C^2 - \sqrt{2}s_0^2 > 0$ for all C.

4.3.4 Subcase (i): $s_0^2 < 1/\sqrt{2}$

Consider the roots (4.50) in this case. For $C = 0$ the roots satisfy

$$s_1(0) = \frac{s_0}{\sqrt{2}} \frac{|s_0^2 - \sqrt{2}| - s_0^2}{1 - \sqrt{2}s_0^2} = \frac{s_0}{\sqrt{2}} \frac{\sqrt{2} - s_0^2 - s_0^2}{1 - \sqrt{2}s_0^2} = s_0,$$

$$s_2(0) = -\frac{s_0}{\sqrt{2}} \frac{|s_0^2 - \sqrt{2}| + s_0^2}{1 - \sqrt{2}s_0^2} = -\frac{s_0}{\sqrt{2}} \frac{\sqrt{2}}{1 - \sqrt{2}s_0^2} = \frac{s_0}{\sqrt{2}s_0^2 - 1} < 0.$$

Note that both $s_1(C) \to 0$ and $s_2(C) \to 0$ as $|C| \to \infty$. Furthermore, one can show that $s_1(C)$ is decreasing and $s_2(C)$ is increasing. Therefore,

$$s_2(C) < 0 < s_1(C) < s_0$$

for all C. Since $s_0^2 < 1/\sqrt{2}$ it is true that $s_0 < \sqrt{2}/s_0$; that is, $s_{\xi\xi} > 0$ (concave up) for all sufficiently small ξ and until the graph of $s(\xi)$ crosses $\sqrt{2}/s_0$. For $C \leq 0$ the graph of $s(\xi)$ is thus always increasing with an asymptotic slope given by $s_\xi^2 = ((1 + C^2)/s_0^2) - \sqrt{2}$ by (4.49). If $C > 0$, the graph decreases to some point ξ_0 where $s(\xi_0) = s_1(C)$ and is increasing thereafter. Figure 4.6 depicts subcase (i) for representative values of C.

It is important to note that this anomalous dispersion case is not materially different from the normal dispersion case. In particular, there is a compression length, an optimal prechirp, and an optimal input pulse width in analogy to the corresponding optima found in section 4.2; these were developed in [BPJ, JaS]. The main point is that if the nonlinearity parameter $s_0 = N$ is sufficiently small, namely, $s_0^2 < 1/\sqrt{2}$, then the pulse width $s(\xi)$ behaves as it would under normal dispersion. Indeed, numerical simulations using the variational method in this case are quite accurate, even with regard to computing optima as in section 4.2 [BPJ, JaS].

4.3.5 Subcase (ii): $(1/\sqrt{2}) < s_0^2 < \sqrt{2}$

Noting the denominators of (4.50) suggests a further division of cases into

(iia) $(1/\sqrt{2}) < s_0^2 < \sqrt{2}$ and $C^2 < \sqrt{2}s_0^2 - 1$, or

(iib) $(1/\sqrt{2}) < s_0^2 < \sqrt{2}$ and $C^2 > \sqrt{2}s_0^2 - 1$.

4.3.6 Subcase (iia): $(1/\sqrt{2}) < s_0^2 < \sqrt{2}$ and $C^2 < \sqrt{2}s_0^2 - 1$

Here $s_2(C) > 0$ in (4.50). Equations (4.49) and (4.50), and the condition $s_1(0) = s_0$, still hold, but now

$$s_2(0) = \frac{s_0}{\sqrt{2}s_0^2 - 1} > 0.$$

In addition,

$$s_0 < \frac{s_0}{\sqrt{2}s_0^2 - 1}$$

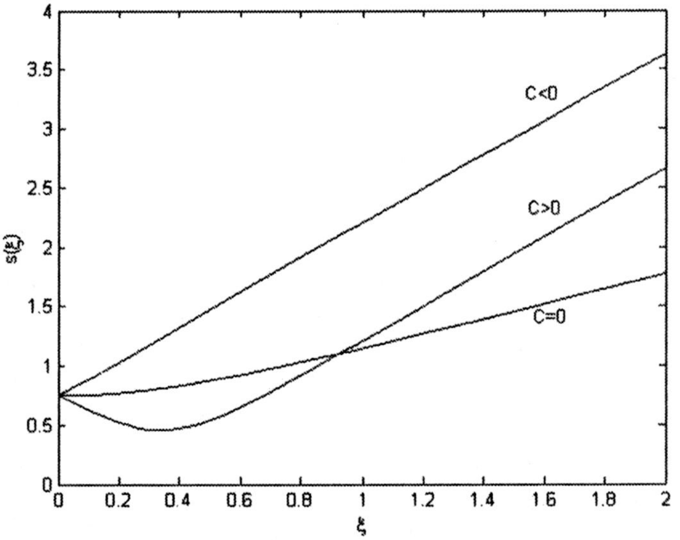

Figure 4.6. *Representative plots of $s(\xi)$ for anomalous dispersion with $s_0 = .75$.*

because this is equivalent to $\sqrt{2}s_0^2 - 1 < 1$, $\sqrt{2}s_0^2 < 2$, or $s_0^2 < \sqrt{2}$, which is assumed in this case. Likewise, $s_0 < (\sqrt{2}/s_0)$ and

$$s_2(0) = \frac{s_0}{\sqrt{2s_0^2 - 1}} > \frac{\sqrt{2}}{s_0}$$

since this is the same as $s_0^2 > 2s_0^2 - \sqrt{2}$ or $s_0^2 < \sqrt{2}$. One can show that $s_1(C)$ is decreasing in C and that $s_2(C)$ is increasing in C until the denominators of (4.50) vanish at the asymptote $C^2 = \sqrt{2}s_0^2 - 1$.

The denominators of (4.50) are negative. The numerator of $s_1(C)$ is negative if and only if $\sqrt{(s_0^2 - \sqrt{2})^2 + 2C^2} < s_0^2$ or $s_0^4 - 2\sqrt{2}s_0^2 + 2 + 2C^2 < s_0^4$ or $C^2 < \sqrt{2}s_0^2 - 1$, which is assumed. Therefore, $s_1(C) > 0$. Putting everything together, one finally has

$$0 < s_1(C) < s_0 < \frac{\sqrt{2}}{s_0} < s_2(C)$$

in case (iia). The graph of $s(\xi)$ is therefore oscillatory, trapped between the critical points $s_1(C)$ and $s_2(C)$. Depending on the sign of C, $s(\xi)$ can be either increasing or decreasing initially but is concave up above the value $\sqrt{2}/s_0$. Figure 4.7 illustrates subcase (iia) for representative values of C.

Subcase (iia) corresponds to the formation of optical solitons. As pointed out by Anderson [And], the first of (4.49) (but not (4.47)),

$$s_{\xi\xi} = \frac{1}{s^3} - \frac{s_0}{\sqrt{2}s^2}, \tag{4.51}$$

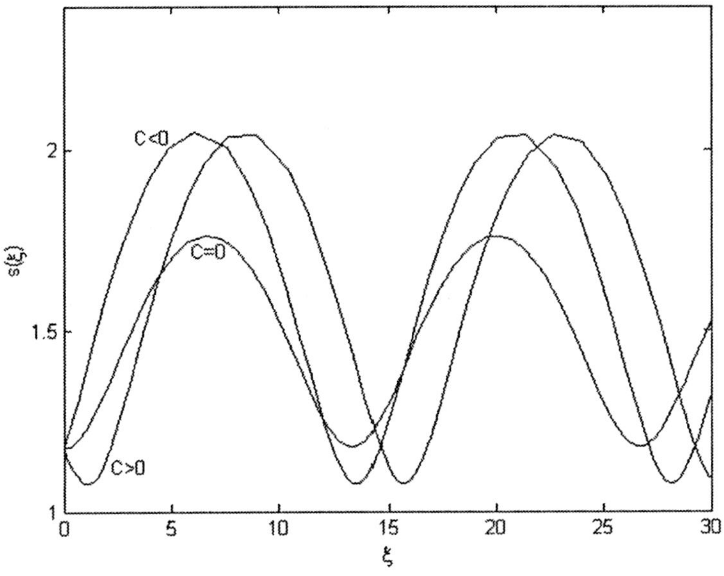

Figure 4.7. *Anomalous dispersion in subcase* (iia) *with* $s_0 = 1.18$.

represents a potential well in which the states are trapped if the well is sufficiently deep. The minus sign in (4.51) arises from the anomalous dispersion condition $\beta_2 < 0$ and causes the nonlinear ($s_0 s^{-2}$) and dispersive (s^{-3}) terms in (4.43) to balance. This constitutes an argument that solitons occur only under anomalous dispersion. However, the method breaks down numerically. As will be discussed in Chapter 5, if $\sqrt{\pi/8} < N < 3\sqrt{\pi/8}$, then the initial Gaussian pulse evolves as $\xi \to \infty$ into the classical hyperbolic secant soliton. Recalling $s_0 = N$, the nonlinearity level $N = 1/\sqrt[4]{2}$ represents a kind of Gaussian-variational threshold for "soliton" formation. It is emphasized that solutions of (3.51) are not periodic in this case; the behavior indicated in Figure 4.7 is artificial and simply serves as an analogy. Indeed, the variational approach in this case could be viewed essentially as a mnemonic device. However, the device gives results that are in the right analogy and are roughly correct.

Just to finish out the analysis, for $C = 0$ the roots $s_1(0) = s_0$ and $s_2(0) = \frac{s_0}{\sqrt{2}s_0^2 - 1}$ coincide at the threshold $s_0^2 = \sqrt{2}$ and one has $s(\xi) = s_0$ for all ξ. This would correspond to a perfect soliton that propagates with no change in shape. At the threshold $s_0^2 = \sqrt{2}$ and for $C^2 < \sqrt{2}s_0^2 - 1 = 1$, one calculates that

$$s_1(C) = \frac{\sqrt{2}}{1 + |C|}, \qquad s_2(C) = \frac{\sqrt{2}}{1 - |C|}.$$

The solution thus oscillates between these roots; the gap closes and traps the constant soliton $s(\xi) \equiv s_0$ solution when $C = 0$. In this case there is perfect balance between dispersion and nonlinearity (and chirp).

4.3.7 Subcase (iib): $(1/\sqrt{2}) < s_0^2 < \sqrt{2}$ and $C^2 > \sqrt{2}s_0^2 - 1$

This is the interesting case in which a launched pulse has sufficient energy (i.e., N is sufficiently large) to evolve into a soliton, but an overchirp prevents actual formation of the soliton. This phenomenon has been thoroughly studied [KS1] and is known to occur. The actual threshold levels of C at which solitons are prevented depend on the initial pulse shape and have been computed only approximately [KS1]. The level $C^2 > \sqrt{2}s_0^2 - 1$ is again artificial but it is interesting that the variational method predicts the existence of a threshold overchirp and provides a rough estimate of its value. The analysis showing $s(\xi) \to \infty$ when $C^2 > \sqrt{2}s_0^2 - 1$ proceeds along the lines given above and will not be given.

The last case, $s_0^2 > \sqrt{2}$, features periodic solutions of (4.9) for $s(\xi)$, which is again artificial. The analogy is that of higher order solitons, which are indeed periodic, but the correct asymptotic forms are much more complicated than those predicted by (4.9). When $s_0^2 > \sqrt{2}$ there is also a threshold overchirp that is, again, predicted by the variational method.

To summarize the section and chapter, the variational approach has been used successfully in numerous investigations [Ag1, BPJ] for analyzing and predicting pulse widths, compensation lengths, optimal chirp, and numerous other characteristics. As for the validity of (4.9), this has never been put on a rigorous basis. In fact, its usefulness is not based at all on the ability to accurately compute solutions to the NLSE. That is well provided by the split step numerical algorithm. Instead, the variational method serves as an ad hoc procedure with an uncanny ability to mimic solutions of (3.1), even in the anomalous dispersion case, and permits symbolic manipulation leading to analysis and optimization.

Chapter 5

Optical Solitons

The discovery and use of the optical soliton effect is one of the most important developments in communications over the last thirty years. Solitons are special types of solutions to the NLSE (3.51) or any one of its equivalent forms such as (3.1) or (3.50). One particular kind of soliton, the fundamental soliton, propagates without change in pulse or spectral amplitude. This is the dream of the communications engineer: a pulse that looks the same coming out of the fiber as it did going in. Moreover, solitons are robust in the sense that an inserted pulse of sufficient energy, but whose shape is not perfect, automatically adjusts itself and evolves during propagation into a perfectly shaped soliton.

However, as mentioned in section 4.3 the NLSE does not include terms accounting for loss, higher order dispersion, or higher order nonlinear effects such as in (3.46). Consequently, optical solitons as described by (3.1) do not actually exist in real fibers. On the other hand, there are ways to partially compensate for loss and other effects so that pulses that behave like solitons over limited propagation lengths are attainable; that is, there is a real and observable *soliton effect*. It is interesting that the optical soliton, which is a mathematical construction, has helped to shape, and sometimes dominate, thinking in the design of optical systems.

The need for researchers in optical communications to understand solitons is fundamental and serves as the motivation for this chapter. The first section surveys the main ideas and results from inverse scattering theory concerning solitons, the purpose being to provide an overview of the field and a road map of the main literature. After that the discussion will move towards some frontiers of research. The second section takes up issues of pulse design for soliton stability, and the third section presents a newly obtained energy threshold for soliton formation. The chapter closes with a few comments in section 5.4. Throughout the chapter, "soliton" will refer to the temporal rather than spatial variable. *Spatial solitons*, which are continuous beams of light trapped spatially by various means [KiSt], will not be considered.

5.1 Background in Solitons for the NLSE

The normalized NLSE in the focusing case is (see section 3.4)

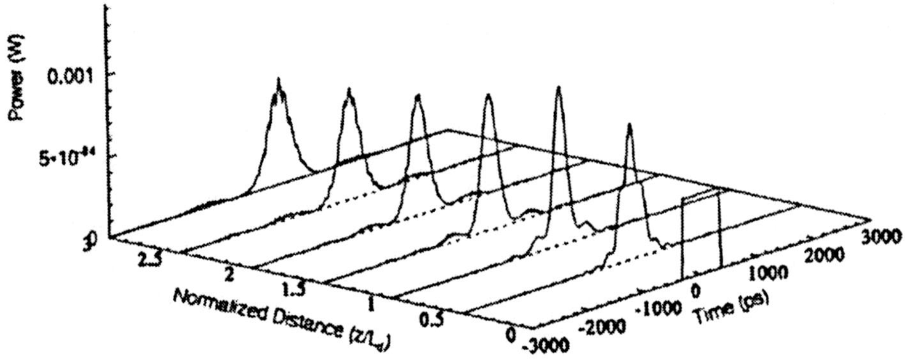

Figure 5.1. *Evolution of a rectangular pulse into a fundamental soliton* [XMa].

$$iu_\zeta + (1/2)u_{\tau\tau} + |u|^2 u = 0. \tag{5.1}$$

By direct substitution one shows that the function $u(\zeta, \tau) = \text{sech}(\tau)e^{i\zeta/2}$ satisfies (5.1). The inserted pulse $u(0, \tau) = \text{sech}(\tau)$ therefore retains its constant shape $|u(\zeta, \tau)| = \text{sech}(\tau)$ exactly for all ζ. Note also that $u(\zeta, \tau) = N\,\text{sech}(\tau)e^{i\zeta/2}$ satisfies (5.1) only if $N = 1$. Likewise, one can show that [Ag1]

$$u(\zeta, \tau) = \frac{4[\cosh(3\tau) + 3\exp(4i\zeta)\cosh(\tau)]\exp(i\zeta/2)}{\cosh(4\tau) + 4\cosh(2\tau) + 3\cos(4\zeta)} \tag{5.2}$$

is a solution of (5.1), whose modulus is periodic in ζ with period $\pi/2$, and has the property that $u(0, \tau) = 2\,\text{sech}(\tau)$. In general it is the case that initial pulses $u(0, \tau) = N\,\text{sech}(\tau)$ produce closed form periodic solutions precisely when N is a positive integer. The $N = 1$ case is called the fundamental soliton, the $N = 2$ case is the second-order soliton, and so forth. It is the fundamental soliton that is of primary interest in communications.

Because of imperfect transmission equipment, it is not possible for an inserted pulse to be an exact hyperbolic secant shape, $u(0, \tau) = \text{sech}(\tau)$, and this raises the question as to what then happens to the evolving solution $u(\zeta, \tau)$ of (5.1). As mentioned, if the input energy lies in a certain range (discussed specifically in section 5.3 below), then the initial pulse evolves into the hyperbolic secant regardless of its particular shape. To illustrate, Figure 5.1 shows the evolution of $|u(\zeta, \tau)|$ over several dispersion lengths, where the input pulse is rectangular with peak power $0.5mW$ and initial width a few hundred ps. As the figure shows, the pulse develops into a hyperbolic secant, overlaid with an oscillation, while excess energy is shed away in the form of side lobes that move farther away from the center of the pulse with increasing distance. The amplitude of the overlaid oscillation eventually decays as $\zeta \to \infty$, leaving only the pure soliton.

This asymptotic behavior, along with how one arrives at $u(\zeta, \tau) = \text{sech}(\tau)e^{i\zeta/2}$ or the form (5.2), is fully explicable only in the context of the *inverse scattering transform* (IST) method, which is the principal method for solving and analyzing (5.1). The IST approach will now be discussed.

To begin with a brief synopsis, IST theory appends to (5.1) a certain linear spectral

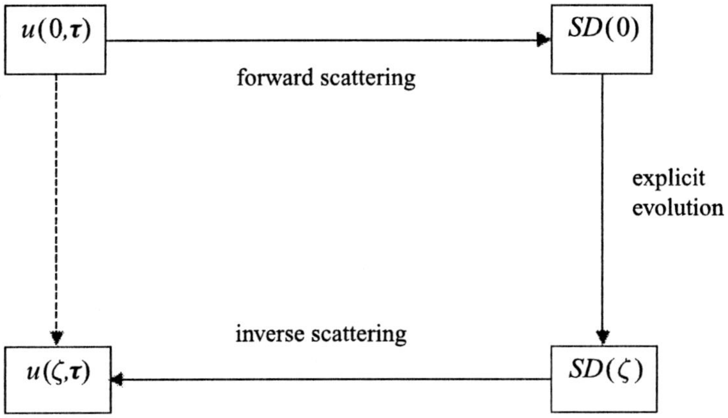

Figure 5.2. *Diagram of the IST approach.*

or eigenvalue problem on the real line for each fixed ζ. The spectral problem has *eigenvalues*, associated *normalization constants*, and a *reflection coefficient* (these will be defined presently), all of which comprise the so-called *scattering data*, denoted by $SD(\zeta)$. The scattering data $SD(0)$ for $\zeta = 0$ is determined solely by $u(0, \tau)$ through *forward scattering theory*. The data $SD(\zeta)$ evolves from $SD(0)$ in an explicit way; this is the essential key to how the IST method works. Finally, $u(\zeta, \tau)$ is determined from $SD(\zeta)$ by *inverse scattering theory*. These steps are explained in some detail below. The diagram in Figure 5.2 appears in most references on IST theory and captures the basic idea.

As will be seen, the explicit evolution of $SD(\zeta)$ is a consequence of the property that the spectral problem appended to (5.1) evolves linearly in ζ, despite that fact that (5.1) is nonlinear. The NLSE is one of a family of equations that share this property. Other members of the family include the Korteweg–de Vries (KdV), modified Korteweg–de Vries (MKdV), and sine-Gordon equations [AbS, BDT]. Because there are explicit techniques for solving these nonlinear equations, at least in principle, the equations are said to be *integrable* by the IST. A number of related integrable equations are considered in [AkA].

The IST method was discovered in 1967 by Gardner et al. [AbS] in the context of the KdV equation $u_\zeta + 6uu_{\tau\tau} + u_{\tau\tau\tau} = 0$. The extension of the IST to the NLSE was obtained by Zakharov and Shabat [ZaS]. The original method was placed into a general operator context by Lax and later formalized by Ablowitz, Kaup, Newell, and Segur (AKNS) [AbS]; the *AKNS procedure* includes both the KdV and NLSE. There are numerous references that cover solitons, the IST, and their histories; some of these are listed at the end of this section.

5.1.1 Zakharov–Shabat Systems

The spectral problem appended to (5.1) is known as a Zakharov–Shabat system and has the form

$$\vec{v}_\tau(\zeta, \tau) = \begin{pmatrix} -i\xi & u(\zeta, \tau) \\ -u(\zeta, \tau)^* & i\xi \end{pmatrix} \vec{v}(\zeta, \tau), \quad -\infty < \tau < \infty, \tag{5.3}$$

where ξ is a complex eigenvalue or spectral parameter and the asterisk denotes the complex conjugate. In (5.3) the normalized distance ζ is regarded as fixed but arbitrary. The solution $\vec{v}$ is a function of three variables, $\vec{v} = \vec{v}(\zeta, \tau, \xi)$, but symbols will often be suppressed to simplify notation. In order to guarantee existence of scattering data of (5.3), $u(\zeta, \tau)$ should satisfy certain integrability and decay conditions as $|\tau| \to \infty$. These will be discussed at the end of the section but, for the moment, assume that the following quantities exist so that they can be defined.

The *eigenvalues* of (5.3) are defined as those complex numbers ξ in the upper half-plane, $\mathrm{Im}(\xi) > 0$, such that (5.3) has a nontrivial solution $\vec{v}(\tau) = \vec{v}(\zeta, \tau, \xi)$ in $L^2(-\infty, \infty)$; that is, $\int_{-\infty}^{\infty} |\vec{v}(\tau)|^2 d\tau < \infty$. It will be convenient to assume that there are finitely many eigenvalues, ζ_m, $m = 1, 2, \ldots, M$; conditions on $u(0, \tau)$ that guarantee a finite number will be given later. The symmetries of (5.3) between upper and lower half-planes allow one to construct a complete spectral theory in the upper half-plane [AbS].

If $u(\zeta, \tau)$ decays sufficiently rapidly as $|\tau| \to \infty$, then there exist solutions $\vec{\psi}(\tau)$, $\vec{\bar{\psi}}(\tau)$, $\vec{\varphi}(\tau)$, and $\vec{\bar{\varphi}}(\tau)$ (all are functions of τ, ξ, and ζ) which satisfy the asymptotic conditions [AbS]

$$\begin{aligned} \vec{\psi}(\tau, \xi) &\cong \begin{pmatrix} 0 \\ 1 \end{pmatrix} e^{i\xi\tau} \quad \text{and} \quad \vec{\bar{\psi}}(\tau, \xi) \cong \begin{pmatrix} 1 \\ 0 \end{pmatrix} e^{-i\xi\tau} \quad \text{as } \tau \to \infty, \\ \vec{\varphi}(\tau, \xi) &\cong \begin{pmatrix} 1 \\ 0 \end{pmatrix} e^{-i\xi\tau} \quad \text{and} \quad \vec{\bar{\varphi}}(\tau, \xi) \cong \begin{pmatrix} 0 \\ -1 \end{pmatrix} e^{i\xi\tau} \quad \text{as } \tau \to -\infty. \end{aligned} \tag{5.4}$$

These solutions of (5.3), which are called *Jost solutions*, are unique up to constant multiples. If ξ is an eigenvalue, then the eigensolution $\vec{v}(\tau, \xi)$ is thus necessarily a multiple of each of $\vec{\varphi}(\tau)$ and $\vec{\psi}(\tau)$. In particular, $\vec{\varphi}(\tau; \xi)$ and $\vec{\psi}(\tau, \xi)$ are linearly dependent at an eigenvalue ξ_m.

In general, $\vec{\varphi}(\tau)$ is a linear combination of $\vec{\psi}(\tau)$ and $\vec{\bar{\psi}}(\tau)$, say

$$\vec{\varphi}(\tau, \xi) = c_{11}(\xi)\vec{\psi}(\tau, \xi) + c_{12}(\xi)\vec{\bar{\psi}}(\tau, \xi), \tag{5.5}$$

and similarly

$$\vec{\psi}(\tau, \xi) = c_{21}(\xi)\vec{\bar{\varphi}}(\tau, \xi) + c_{22}(\xi)\vec{\varphi}(\tau, \xi). \tag{5.6}$$

The scattering data of (5.3) is built from the coefficients $c_{mn}(\xi)$in (5.5) and (5.6). The *transmission coefficient* $T(\xi)$ is defined by $T(\xi) = 1/c_{12}(\xi)$, which can be shown to equal $-1/c_{21}(\xi)$ using Wronskians [AbS]. The poles of $T(\xi)$ are the eigenvalues of (5.3) [AbS]. For each eigenvalue ξ_m there is a *normalization constant* C_m given by $C_m = c_{11}(\xi_m)/\dot{c}_{12}(\xi_m)$, where the overdot denotes a derivative with respect to ξ, which links the Jost solutions at $\xi = \xi_m$ by $\vec{\varphi}(\tau, \xi_m) = C_m \vec{\psi}(\tau, \xi_m)$. Finally, in addition to the transmission coefficient there is a *reflection coefficient* defined by $R(\xi) = c_{11}(\xi)/c_{12}(\xi)$. Here $R(\xi)$ is actually the *right reflection coefficient* for (5.3); there is also a *left coefficient*, but it duplicates the information embedded in $R(\xi)$ [Lam]. The eigenvalues, norming constants, and reflection coefficient comprise the *scattering data*,

$$SD(\zeta) = \left\{ \{\xi_m\}_{m=1}^{M}, \{C_m\}_{m=1}^{M}, R(\xi) \right\}. \tag{5.7}$$

All quantities in (5.7) depend on the distance variable ζ. For fixed ζ the problem of finding the quantities (5.7) from (5.3) is called the *forward scattering problem.*

In particular, the top part of the diagram in Figure 5.2, moving left to right, consists of setting $\zeta = 0$ in (5.1) and calculating $SD(0)$ in (5.7).

The data (5.7) is *complete* for each ζ in the sense that knowledge of the eigenvalues, norming constants, and reflection coefficient uniquely determine $u(\zeta, \tau)$. There is an algorithm for constructing $u(\zeta, \tau)$ from (5.7), which is straightforward to describe.

5.1.2 Inverse Scattering for Zakharov–Shabat Systems

One can show [AbS] that there exists a *transformation kernel*

$$\vec{K}(\tau, s) = \begin{pmatrix} K_1(\tau, s) \\ K_2(\tau, s) \end{pmatrix}$$

such that the Jost solution (5.6) has the representation

$$\vec{\psi}(\tau, \xi) = \begin{pmatrix} 0 \\ 1 \end{pmatrix} e^{i\xi\tau} + \int_\tau^\infty \vec{K}(\tau, s) e^{i\xi s} ds.$$

In fact, given the scattering data (5.7), $\vec{K}(\tau, s)$ is obtained by first defining

$$F(\tau) = \frac{1}{2\pi} \int_{-\infty}^{\infty} R(\xi) e^{i\xi\tau} d\xi - i \sum_{m=1}^{M} C_m e^{i\xi_m \tau}$$

and then solving the integral equation

$$\begin{pmatrix} K_2^*(\tau, y) \\ -K_1^*(\tau, y) \end{pmatrix} + \begin{pmatrix} 0 \\ 1 \end{pmatrix} F(\tau + y) + \int_\tau^\infty \vec{K}(\tau, s) F(s + y) ds = 0 \tag{5.8}$$

for $\vec{K}(\tau, y)$ [AbS]. Writing $\vec{K}(\tau, s) = \vec{K}(\tau, s, \zeta)$ to denote the dependence on the distance variable ζ, the potential $u(\zeta, \tau)$ is obtained from the kernel by [AbS, Lam]

$$u(\zeta, \tau) = -K_1(\tau, \tau, \zeta). \tag{5.9}$$

Thus (5.8) and (5.9) provide the means of moving right to left across the bottom of the diagram in Figure 5.2. As mentioned, finding $SD(0)$ in (5.7) directly from $u(0, \tau)$ in (5.3) takes one left to right across the top of the diagram. What remains is to explain how $SD(0)$ evolves into $SD(\zeta)$, that is, going from top to bottom on the right side of the diagram. The most direct way to see this is to sketch the AKNS procedure. This will now be done through a series of formal or nonrigorous calculations, the purpose being to link the system (5.3) directly to the NLSE (5.1). That is, the formal derivation will show that the Zakharov–Shabat system (5.3) is an appropriate mechanism for describing the evolution of $u(\zeta, \tau)$ in the ζ variable.

5.1.3 Linkage of the NLSE with Zakharov–Shabat Systems

Start with some given function $u(\zeta,\tau)$ and form the system (5.3) in which ζ is fixed but arbitrary. The basic notion in the IST approach is to build the evolution of the scattering data $SD(\zeta)$ around the key idea of holding the eigenvalues fixed as functions of ζ. That is, the quantities $\xi_m(\zeta)$ are going to be constant functions of ζ for each m. It turns out that this condition, coupled with postulating that $\vec{v}(\tau,\xi,\zeta)$ evolves linearly in ζ, also implies a very convenient pattern of evolution of the norming constants and reflection coefficient in (5.7). To proceed with this, assume that $\vec{v}(\tau,\xi,\zeta)$ evolves linearly as a function of ζ and in such a way that the eigenvalues $\xi(\zeta)$ are stationary, that is,

$$\vec{v}_\zeta = \begin{pmatrix} A & B \\ C & D \end{pmatrix} \vec{v}, \qquad \xi_\zeta = \partial\xi/\partial\zeta = 0, \tag{5.10}$$

where the entries of the matrix are functions of ζ, τ, and ξ. Take the ζ partial (∂_ζ) of (5.3), the τ partial (∂_τ) of (5.10), equate the two mixed partials, and set $\xi_\zeta = 0$. The result is the set of equations [AbS]

$$\begin{aligned} A_\tau &= Cu + Bu^*, \\ B_\tau + 2i\xi B &= -Au + Du + u_\zeta, \\ C_\tau - 2i\xi C &= -Au^* + Du^* - u_\zeta^*, \\ D_\tau &= -Cu - Bu^*. \end{aligned} \tag{5.11}$$

The first and last equations of (5.11) imply that $A_\tau = -D_\tau$ or $A = -D + \eta(\zeta,\xi)$, with $\eta(\zeta,\xi)$ being an arbitrary function. Taking $\eta(\zeta,\xi) = 0$ gives $A = -D$ so that (5.11) reduces to

$$\begin{aligned} A_\tau &= Cu + Bu^*, \\ B_\tau + 2i\xi B &= -2Au + u_\zeta, \\ C_\tau - 2i\xi C &= -2Au^* - u_\zeta^*. \end{aligned} \tag{5.12}$$

Now seek solutions to (5.12) in the form

$$\begin{aligned} A &= A_0 + A_1\xi + A_2\xi^2, \\ B &= B_0 + B_1\xi + B_2\xi^2, \\ C &= C_0 + C_1\xi + C_2\xi^2. \end{aligned} \tag{5.13}$$

Substitute (5.13) into (5.12) and equate coefficients of powers of ξ. For example, from the second equation of (5.12), equating the coefficients of ξ^0 gives

$$(B_0)_\tau = -2uA_0 + u_\zeta. \tag{5.14}$$

Similarly, other operations lead to

$$B_2 = 0, \qquad C_2 = 0, \qquad (A_2)_\tau = 0,$$

so that one may take $A_2 = a_2(\zeta)$ for some function a_2. Further conclusions are [AbS]

$$
\begin{aligned}
&B_1 = ia_2(\zeta)u, \qquad C_1 = -ia_2(\zeta)u^*, \qquad A_1 = a_1(\zeta), \\
&A_0 = -\frac{1}{2}a_2(\zeta)|u|^2, \qquad B_0 = -ia_1(\zeta)u - \frac{1}{2}a_2(\zeta)u_\tau, \\
&C_0 = -ia_1(\zeta)u^* - \frac{1}{2}a_2(\zeta)u_\tau^*,
\end{aligned} \tag{5.15}
$$

where a_1 and a_2 are arbitrary functions. Choose $a_1 = 0$ and $a_2 = -2i$, substitute these into the expression for B_0 in (5.15), and put the result into (5.13) to obtain

$$\left(-\frac{1}{2}[-2i]u_\tau\right)_\tau = -2A_0 + u_\zeta.$$

From the expression for A_0 in (5.15) one gets $iu_{\tau\tau} = -2i|u|^2u + u_\zeta$ and finally

$$iu_\zeta + u_{\tau\tau} + 2|u|^2u = 0. \tag{5.16}$$

Equation (5.16) is equivalent to (5.1), as seen by setting $\tilde{\zeta} = \frac{1}{2}\zeta$ in (5.1) and then computing $iu_{\tilde{\zeta}} + u_{\tau\tau} + 2|u|^2u = 0$.

Thus the NLSE is linked to Zakharov–Shabat systems at least by formal algebraic manipulations. These manipulations show that (5.10), (5.3), and (5.1) are consistent, meaning that there are choices for the entries of the matrix

$$\Gamma = \begin{pmatrix} A & B \\ C & D \end{pmatrix}$$

in (5.10) such that the eigenvalues of (5.3) are fixed as functions of ζ when $u(\zeta,\tau)$ satisfies (5.1). Collecting the various terms in (5.13), (5.10) becomes

$$\vec{v}_\zeta = \begin{pmatrix} i(|u|^2 - 2\xi^2) & iu_\tau + 2\xi u \\ iu_\tau^* - 2\xi u^* & -i(|u|^2 - 2\xi^2) \end{pmatrix} \vec{v}. \tag{5.17}$$

Some remarks on making the AKNS formalism rigorous will be given below. For now, however, system (5.17) can be used to determine the evolution of the norming constants and reflection coefficient in (5.7).

5.1.4 Evolution of the Scattering Data as Functions of ζ

Suppose $\vec{v}(\tau)$ is an eigenfunction of (5.3), so that $\vec{v}(\tau)$ is a multiple of the Jost solution $\vec{\psi}(\tau)$. The multiplicative factor, say $f(\zeta)$, can be chosen so that $\vec{v}(\zeta,\tau,\xi) = f(\zeta)\vec{\psi}(\zeta,\tau,\xi)$ satisfies (5.17), where ξ is an eigenvalue. In fact, substituting $\vec{v}(\zeta,\tau,\xi) = f(\zeta)\vec{\psi}(\zeta,\tau,\xi)$ into (5.17) and setting $\partial\xi/\partial\zeta = 0$ gives

$$
\begin{aligned}
\partial_\zeta(v_1) &= f'(\zeta)\psi_1 + f(\zeta)\partial_\zeta(\psi_1) \\
&= i(|u|^2 - 2\xi^2)f(\zeta)\psi_1 + (iu_\tau + 2\xi u)f(\zeta)\psi_2,
\end{aligned} \tag{5.18}
$$

which can be viewed as an ordinary differential equation in the unknown $f(\zeta)$. Incidentally, the second component of (5.17) is redundant with (5.18). Let $\tau \to +\infty$ in (5.18). By

(5.4) $v_2 \cong f(\zeta)e^{i\xi\tau}$ and by (5.17) $\partial_\zeta(v_2) \cong 2i\xi^2 v_2$ since it is assumed that $u, u_\tau \to 0$. Formally differentiating $v_2 \cong f(\zeta)e^{i\xi\tau}$ with respect to ζ and using $\partial_\zeta(v_2) \cong 2i\xi^2 v_2$ gives $f'(\zeta)e^{i\xi\tau} \cong 2i\xi^2 f(\zeta)e^{i\xi\tau}$ for large τ. However, this is independent of τ and so one can solve the resulting differential equation to obtain

$$f(\zeta) = f(0)e^{2i\xi^2\zeta}. \tag{5.19}$$

Now let $\tau \to -\infty$. By (5.6), (5.4), and (5.19)

$$\begin{aligned}\vec{v}(\zeta,\tau,\xi) &= f(\zeta)\vec{\psi}(\zeta,\tau,\xi) = f(0)e^{2i\xi^2\zeta}[c_{21}\vec{\bar{\varphi}} + c_{22}\vec{\varphi}] \\ &\cong f(0)e^{2i\xi^2\zeta}\left[c_{21}\begin{pmatrix} 0 \\ -1 \end{pmatrix} e^{i\xi\tau} + c_{22}\begin{pmatrix} 1 \\ 0 \end{pmatrix} e^{-i\xi\tau}\right],\end{aligned}$$

and consequently

$$v_1 \cong f(0)e^{2i\xi^2\zeta}c_{22}e^{-i\xi\tau}, \qquad v_2 \cong -f(0)e^{2i\xi^2\zeta}c_{21}e^{i\xi\tau}. \tag{5.20}$$

Differentiating the first of (5.20) gives

$$\partial_\zeta(v_1) \cong f(0)e^{2i\xi^2\zeta}e^{-i\xi\tau}\left[\partial_\zeta(c_{22}) + 2i\xi^2 c_{22}\right] \tag{5.21}$$

and, substituting from (5.17) with the assumption $u, u_\tau \to 0$, this expression becomes

$$\partial_\zeta(v_1) \cong -2i\xi^2 v_1 \cong -2i\xi^2 f(0)e^{2i\xi^2\zeta}c_{22}e^{-i\xi\tau}.$$

The approximation (5.21) thus becomes

$$\partial_\zeta(c_{22}) + 2i\xi^2 c_{22} \cong -2i\xi^2 c_{22},$$

that is,

$$c_{22} = c_{22}(0)e^{-4i\xi^2\zeta}.$$

Working with v_2 instead of v_1 in (5.20), one obtains

$$\partial_\zeta(v_2) \cong -f(0)e^{2i\xi^2\zeta}e^{i\xi\tau}[\partial_\zeta(c_{21}) + 2i\xi^2 c_{21}],$$

in which $\partial_\zeta(v_2) \cong 2i\xi^2 v_2$ from (5.17). Replacing v_2 by its value in (5.20), there follows

$$\partial_\zeta(c_{21}) + 2i\xi^2 c_{21} \cong 2i\xi^2 c_{21},$$

from which one gets $\partial_\zeta(c_{21}) = 0$ or

$$c_{21} = c_{21}(0);$$

that is, $c_{21}(\zeta)$is constant in ζ. Using the symmetries [Lam] $c_{22}(\xi) = c_{11}(-\xi)$ and $c_{12}(\xi) = -c_{21}(\xi)$ (in the ξ variable) together with the definition $R(\xi) = c_{11}(\xi)/c_{12}(\xi)$ (below (5.6)) gives the evolution of the reflection coefficient as [AbS, Lam]

$$R(\xi,\zeta) = R(\xi,0)e^{-4i\xi^2\zeta}. \tag{5.22}$$

Since c_{12} is independent of ζ, so is $\dot{c}_{12}$, and thus the normalization constants $C_m = c_{11}(\xi_m)/\dot{c}_{12}(\xi_m)$ evolve according to

$$C_m(\zeta) = C_m(0)e^{-4i\xi_m^2\zeta}. \tag{5.23}$$

This completes the description of the dependence of the scattering data (5.7) on the normalized distance ζ.

5.1.5 Solitons and the Asymptotics of $u(\zeta, \tau)$ for Large ζ

Looking back at Figure 5.2, explicit ways have now been described to perform the calculations corresponding to the top, right, and bottom sides of the diagram.

However, the machinery leading from $u(0, \tau)$ to $u(\zeta, \tau)$ has turned out to be more useful for theoretical than practical purposes simply because of the formulaic complexities. The only instance in which $u(\zeta, \tau)$ can be found in explicit form is the *reflectionless* case $R(\xi, 0) = 0$, where one also has $R(\xi, \zeta) = 0$ by (5.22). Indeed, only the single eigenvalue case, $M = 1$, is truly simple, and even this requires some work. To briefly mention this case, suppose that (5.3) for $\zeta = 0$ has a single eigenvalue, $\xi = \xi_1$, and $R(\xi, 0) = 0$. In (5.8) one then has $F(\tau) = C_1(0)e^{i\xi_1\tau}$, where $C_1(0)$ is the initial norming constant. Some manipulations [AbS] show that the transformation kernel in (5.8) has first component

$$K_1(\tau, y, \zeta) = iC_1(\zeta)^* e^{-i\xi_1^*(\tau+y)}\left[1 - \frac{|C_1(\zeta)|^2}{(\xi_1 - \xi_1^*)}e^{2i(\xi_1-\xi_1^*)\tau}\right], \tag{5.24}$$

where $C_1(\zeta)$ is given by (5.23). Considering the generic case where $\xi_1 = i\eta_1$ is purely imaginary (see section 5.2), and setting $y = \tau$ in (5.24), (5.9) becomes

$$u(\zeta, \tau) = 2\eta_1 e^{-4i\eta_1^2\zeta} e^{ik_0} \operatorname{sech}(2\eta_1\tau - \tau_0), \tag{5.25}$$

where k_0 and τ_0 are constants. Among other things (5.25) shows that the only pulse shape having a purely imaginary eigenvalue and vanishing reflection coefficient is the hyperbolic secant. To derive the analog of (5.25) when there are two eigenvalues is significantly more involved [NMPZ]. A prototype case of two eigenvalues is discussed at the beginning of section 5.2.

Perhaps the most significant application of the IST method is in information provided by the asymptotics of $u(\zeta, \tau)$ for large ζ. The IST approach provides a decomposition of $u(\zeta, \tau)$ into the sum of a stable soliton part and a dispersive remainder of the form [NMPZ, Kot]

$$u(\zeta, \tau) = \sum_{m=1}^{M} u_m(\zeta, \tau) + u_D(\zeta, \tau), \tag{5.26}$$

where the finite sum corresponds to the eigenvalues of the scattering data (5.7) and the dispersive term $u_D(\zeta, \tau)$ decays for fixed τ as $\zeta \to \infty$. In (5.26) it is assumed that there are finitely many eigenvalues. To illustrate (5.26) consider a result of Kotlyarov [Kot], which assumes that the problem

$$iu_\zeta + (1/2)u_{\tau\tau} + |u|^2 u = 0, \qquad u(0, \tau) \text{ given}, \tag{5.27}$$

has a smooth unique solution $u(\zeta, \tau)$ such that the scattering data (5.7) exists with M eigenvalues $\xi_m = \omega_m + i\eta_m$, $m = 1, \ldots, M$. Then $u(\zeta, \tau)$ can be represented as

$$u(\zeta, \tau) = \sum_{m=1}^{M} 2\eta_m \frac{e^{-2i\omega_m\tau - 4i(\omega_m^2 - \eta_m^2)\zeta - i\delta_m}}{\cosh[2\eta_m(\tau + 4\omega_m\zeta - \tau_m)]} + u_D(\zeta, \tau), \tag{5.28}$$

where δ_m and τ_m are constants and the dispersive part $u_D(\zeta, \tau) \to 0$ as $\zeta \to \infty$. As is clear from (5.24) and (5.25), the norming constants $C_m(\zeta)$ given by (5.23) are involved in the coefficients in (5.28). In the generic case where $\omega_m = 0$ (purely imaginary eigenvalues; see section 5.2) the finite sum, or discrete part, of (5.28) has constant modulus with $M = 1$ and is periodic in ζ for $M > 1$ [Ag1].

It can be shown that the decay rate of $u_D(\zeta, \tau)$ has order of magnitude $\zeta^{-1/2}$ [AbS, NMPZ]. Thus, with the representation (5.28), one can encapsulate pulse behavior for (5.1). An input pulse $u(0, \tau)$ determines associated initial scattering data $SD(0)$ in (5.7). The eigenvalues ξ_m do not change with increasing ζ and completely determine, along with the norming constants $C_m(\zeta)$, the long distance behavior of $u(\zeta, \tau)$. For large ζ the pulse approaches, at the rate $\zeta^{-1/2}$, a finite sum of terms that are explicit multiples of hyperbolic secants. Therefore, the pulse evolves into superimposed hyperbolic secants irrespective of the initial shape as long as there are eigenvalues. If there are no eigenvalues, then (5.28) shows that $u(\zeta, \tau)$ is purely dispersive; that is, $u(\zeta, \tau) \to 0$.

The existence, number, and location of eigenvalues will be discussed in more detail in sections 5.2 and 5.3.

5.1.6 Well-Posedness of Problem (5.27)

Hitherto all of the discussions in this chapter have been heuristic. Conditions on $u(0, \tau)$ that guarantee existence of scattering data $SD(0)$, the finiteness of the number of eigenvalues, and other important conditions have been omitted in the interests of simple exposition. For example, the AKNS formalism indicates symbolically how $SD(\zeta)$ should evolve but does not show that the scattering data itself actually exists. The representation (5.9) is thus completely formal at this point. This essentially describes the state of IST theory from its discovery until the 1980s, when the first rigorous investigations began. The survey article of Xin Zhou [Zh1] contains a historical perspective on the IST, a number of references to other expository material, and an up-to-date account of analytic results.

Beals and Coifman placed the IST on a solid mathematical basis in a set of papers in 1984 and 1985 [BC1, BC2]. The goal of [BC2] was to identify a class of initial values $u(0, \tau)$ for which the AKNS formalism could be completely justified. The starting point for Beals and Coifman is the *Schwartz class* of functions

$$S = \{q \in C^\infty(-\infty, \infty) : \tau^\beta \partial_\tau^\alpha(q) \text{ is bounded for every } \alpha, \beta \geq 0\}.$$

They define a subset $S_0 \subset S$ of *generic potentials* q, whose main property is that the scattering data contains finitely many eigenvalues, and they prove that if $u(0, \tau)$ is in S_0, then (5.1) has a unique solution $u(\zeta, \tau)$ in S for all $\zeta \geq 0$ and that the scattering data $SD(\zeta)$ exists for all $\zeta \geq 0$ and evolves according to AKNS.

The genericity assumption of [BC2] was removed by Zhou [Zh2], who established the well-posedness of (5.27) within a weighted Sobolev space. *Well-posedness* of (5.27) within a class Q will mean here that if initial values $u(0, \tau)$ are in Q, then there is a unique solution $u(\zeta, \tau)$ to (5.1) belonging to Q for every $\zeta \geq 0$ such that the scattering data $SD(\zeta)$ exists, and evolves according to AKNS, and that $u(\zeta, \tau)$ is recoverable uniquely from $SD(\zeta)$. In [Zh2] well-posedness is proved in the class Q consisting of functions $u(\zeta, \tau)$ such that for some $k \geq 1$ the expression $(1 + |\tau|^k)|u(\zeta, \tau)|$ is both integrable and square integrable with respect to τ over $-\infty < \tau < \infty$ for all $\zeta \geq 0$. Zhou proves that the relationships $u(0, \tau) \leftrightarrow SD(0) \leftrightarrow SD(\zeta) \leftrightarrow u(\zeta, \tau)$ are all one-to-one. Even though there can be infinitely many eigenvalues in this case [Zh2], the formula that recovers $u(\zeta, \tau)$ from the scattering data still works because it is set in the framework of a *Riemann–Hilbert problem*. To put this in context, note that the IST machinery described above is problematic, beginning with the definition of $F(\tau)$ in (5.8) and all the way through (5.28), if there are infinitely many eigenvalues. Zhou also gives an example of a Schwartz class potential $u(\zeta, \tau)$ for which there are infinitely many eigenvalues. It is thus difficult to say in general whether the number of eigenvalues of (5.3) is finite.

However, in soliton communication systems the number of eigenvalues of (5.3) must be not only finite but precisely one; that is, only the fundamental soliton can propagate. The second-order soliton (5.2), for example, is periodic in such a way that the field intensity at the pulse center $\tau = 0$ vanishes halfway through the period, making detection impossible if the pulse is sampled at its center (which is the logical place). A precise criterion on the number of eigenvalues in physically realistic systems will be given in section 5.3.

To comment on the IST literature, the AKNS procedure originated in [AKNS] and was covered in detail in [AbS]; other excellent sources are [EcV] and [Sch]. There is also a variational approximation for the IST [Mal]. A three-component analogy of the Zakharov–Shabat system was used in [MBES] to develop an IST for the Manakov equations. This system also comes up in the study of Bose–Einstein condensates [LiC]. In the normal dispersion defocusing case of (3.51), there are no eigenvalues [AbS] and the unchirped solutions of the NLSE are dominated by dispersion. It is of interest in this case to know the long-distance (large ζ) asymptotics of $u(\zeta, \tau)$ in order to predict when pulse intensity falls to undetectable levels as a result of dispersion. The best results in this direction are due to Deift and Zhou [DeZ], who obtained a decay rate of $\zeta^{-1/2}$ with rigorous error bound for initial pulse shapes belonging to a weighted Sobolev space. Let $q(\tau) = u(0, \tau)$ belong to $H^{(1,1)} = \{q : q, q', \tau q(\tau) \in L^2(-\infty, \infty)\}$. The weighted Sobolev space $H^{(1,1)}$ is the largest space in which the Cauchy problem $iu_\zeta - (1/2)u_{\tau\tau} + |u|^2 u = 0$, $u(0, \tau) = q(\tau)$, is well posed and the scattering data exists [DeZ]. Let κ be a constant, $0 < \kappa < 1/4$. Then as $\zeta \to \infty$,

$$u(\zeta, \tau) = k_1 \zeta^{-1/2} e^{i(\tau^2/4\zeta) - k_2 \log(2\zeta)} + O(\tau^{-1/2-\kappa}),$$

where k_1 and k_2 are constants; this bound is uniform for all τ [DeZ]. It would be of interest to extend this asymptotic result to the dispersive part of an expansion of the form (5.28) for the focusing case.

5.2 Purely Imaginary Eigenvalues

This section and the next will consider the eigenvalues of (5.3), where $u(\zeta, \tau)$ is a solution to (5.1). For purposes of counting and locating eigenvalues, it is enough to set $\zeta = 0$ in (5.3) since the eigenvalues are stationary with respect to ζ. To this end it will simplify notation to put $q(\tau) = u(0, \tau)$ and replace (5.3) with

$$\vec{v}'(\tau) = \begin{pmatrix} -i\xi & q(\tau) \\ -q(\tau)^* & i\xi \end{pmatrix} \vec{v}(\tau), \quad -\infty < \tau < \infty, \tag{5.29}$$

where the prime in (5.29) denotes differentiation with respect to τ. The function $q(\tau)$ is often called a *potential* for the Zakharov–Shabat system (5.28). In the remainder of the chapter the number and location of the eigenvalues of (5.29) will be taken up. This section will focus on location.

As mentioned in connection with (5.24) and (5.28), the generic case is the one in which the eigenvalues of (5.29) are purely imaginary. What this means is that if $q(\tau)$ is sufficiently simple and physically plausible, then the eigenvalues are finite in number and lie on the imaginary axis. This is true for hyperbolic secants, Gaussians, rectangular pulses, and other elementary symmetric shapes [KS1]. To see the effects of nonimaginary eigenvalues, consider (5.28) with, for example, two eigenvalues $\xi_m = \omega_m + i\eta_m$, $m = 1, 2$. For the standard symmetric pulse shapes [KS1] any eigenvalues that are not imaginary occur in pairs symmetric about the imaginary axis, that is, $\omega_1 = -\omega_2$ and $\eta_1 = \eta_2$. Assuming this configuration, the terms $\text{sech}[2\eta_1(\tau + 4\omega_1\zeta - \tau_1)]$ and $\text{sech}[2\eta_1(\tau - 4\omega_1\zeta - \tau_2)]$ in (5.28) are traveling waves that walk away from each other on the τ axis with increasing ζ, representing the breakup of an initial pulse. On the other hand, two purely imaginary eigenvalues, $i\eta_1$ and $i\eta_2$, combine to form a second-order periodic soliton.

This raises the issue of when the eigenvalues of (5.29) are on the imaginary axis, a question that has an interesting history. As pointed out in [KS2], for more than 20 years there seems to have been a "folklore theorem" in the literature that the eigenvalues of (2.8) are purely imaginary if $q(\tau)$ is real and symmetric. Although this is true for the simple examples, it is false in general; there are straightforward examples of symmetric step potentials with nonimaginary eigenvalues. There are examples of piecewise constant curves with two lobes, or humps, or concentrations of energy. The principal result of [KS2] is that eigenvalues are purely imaginary when $q(\tau)$ is *single lobe*; that is, $q(\tau) \geq 0$ is increasing on the left of $\tau = 0$ and decreasing to the right. There is no loss of generality in taking $\tau = 0$ to be the center of concentration of energy since the eigenvalues of (5.29) do not change under a shift in the τ variable. Specifically, suppose that $q(\tau)$ is real, nonnegative, bounded, piecewise smooth, absolutely integrable on the real line, and single lobe. Then the eigenvalues are purely imaginary. A sketch of the proof will now be given; details are in [KS2].

5.2.1 Single Lobe Potentials

As a preliminary first write (5.29) in the form

$$iv_1' - iq(\tau)v_2 = \xi v_1, iv_2' + iq(\tau)v_1 = -\xi v_2, \tag{5.30}$$

where ξ is an eigenvalue. Multiply the first equation of (5.30) by v_2^*, the second by v_1^*, and subtract to obtain

$$i(v_1'v_2^* - v_2'v_1^*) - iq(\tau)(|v_1|^2 + |v_2|^2) = \xi(v_1v_2^* + v_2v_1^*). \tag{5.31}$$

Integrating (5.31) over the real axis gives

$$i\int_{-\infty}^{\infty}(v_1'v_2^* - v_2'v_1^*)d\tau - i\int_{-\infty}^{\infty}q(\tau)(|v_1|^2 + |v_2|^2)d\tau = \xi\int_{-\infty}^{\infty}(v_1v_2^* + v_2v_1^*)d\tau. \tag{5.32}$$

Integrating the first integral of (5.32) by parts and using the fact that $|v_1|, |v_2| \to 0, |\tau| \to \infty$, one sees that the first integral in (5.32) is the same as $(\int_{-\infty}^{\infty}(v_1'v_2^* - v_2'v_1^*)d\tau)^*$, its own conjugate. Since the integral is therefore real, the first term in (5.32) must be purely imaginary. The second term in (5.32) is also imaginary since $q(\tau)$ is real. Thus the left side of (5.32) is imaginary. Note that the integral on the right side of (5.32) is real valued, and therefore if this integral is not zero, then ξ must be imaginary. In other words, a criterion that ξ be pure imaginary is that

$$\int_{-\infty}^{\infty} (v_1 v_2^* + v_2 v_1^*) d\tau \neq 0. \tag{5.33}$$

Now let $\xi = \alpha + i\beta$, $\beta > 0$. Using (5.33) it will be shown that $\alpha = 0$. To simplify the argument suppose that $q(\tau)$ is positive and everywhere differentiable; the proof in [KS2] is valid for potentials $q(\tau)$ that are piecewise smooth. Multiply the first equation of (5.30) by v_1^* to obtain $v_2 v_1^* = (v_1' v_1^* / q(\tau)) + (i\xi / q(\tau))|v_1|^2$. If one adds this expression to its conjugate and integrates over an interval $[0, d]$, then

$$\int_0^d (v_2 v_1^* + v_1 v_2^*) d\tau = \int_0^d \frac{v_1' v_1^* + v_1'^* v_1}{q(\tau)} d\tau + i(\xi - \xi^*) \int_0^d \frac{|v_1|^2}{q(\tau)} d\tau. \tag{5.34}$$

The first integral on the right side of (5.34), which equals $\int_0^d \{(v_1 v_1^*)' / q(\tau)\} d\tau$, can be integrated by parts as

$$\int_0^d \{(v_1 v_1^*)' / q(\tau)\} d\tau = \left[\frac{|v_1(s)|^2}{q(s)} \right]_{s=0}^d + \int_0^d \frac{|v_1(\tau)|^2 q'(\tau)}{q(\tau)^2} d\tau. \tag{5.35}$$

It is shown by a technical argument in [KS2] that $\{|v_1(d)|^2 / q(d)\} \to 0, d \to \infty$. Using this and letting $d \to \infty$ in (5.35) gives

$$\int_0^\infty (v_2 v_1^* + v_1 v_2^*) d\tau = -\frac{|v_1(0)|^2}{q(0)} + \int_0^\infty \frac{|v_1(\tau)|^2 q'(\tau)}{q(\tau)^2} d\tau - 2\beta \int_0^\infty \frac{|v_1(\tau)|^2}{q(\tau)} d\tau, \tag{5.36}$$

recalling $\xi = \alpha + i\beta$, $\beta > 0$. Since $q(\tau)$ is decreasing on $[0, \infty)$, then the right side of (5.36) is negative. A similar argument shows that $\int_{-\infty}^0 (v_1 v_2^* + v_2 v_1^*) d\tau < 0$. Coupled with (5.36) this establishes (5.33) and thus shows that ξ is purely imaginary.

The example $q(\tau) = 145.85$, $1.06 \leq |\tau| \leq 1.07$, $q(\tau) = 0$ elsewhere, has an eigenvalue at approximately $\xi = 1.4426 + i$, as cited in [KS2].

5.2.2 Complex Potentials

Complex-valued potentials $q(\tau)$ are also important and were encountered in Chapters 3 and 4 in the context of initial chirp. As mentioned, initial Gaussian pulse shapes of the form (3.40), where C is the chirp parameter, are often used in practice. Recall from section 4.2 that initial chirp is useful in dispersion management and that C can be optimized with respect to output pulse width. For example, there is a unique smallest value of C, $C > 0$, such that the input and output pulse widths are the same after a pulse propagates a distance ζ in the anomalous dispersion (soliton) regime; see the discussion in section 4.2.4.

In fact, it makes sense to consider input pulses of the form (3.40), that is,

$$q(\tau) = u(0, \tau) = q_0(\tau)e^{iC\tau^2}, \tag{5.37}$$

where $q_0(\tau) \geq 0$. A number of studies have tracked the eigenvalues of (5.29), with potentials (5.37), as functions of the parameter C. The article [KS1] surveyed the literature on this problem and investigated the eigenvalue trajectories in detail for various single lobe envelopes $q_0(\tau)$, including rectangular, Gaussian, and hyperbolic shapes. For most practical input pulse shapes [KS1], essentially the same thing happens. When $C = 0$ the eigenvalues (if any) are placed on the imaginary axis, as one knows from the first part of this section. As C is allowed to increase, the eigenvalues vary but are initially confined to the imaginary axis even though (5.37) is complex valued. Depending on particulars of the shape $q_0(\tau)$, either the eigenvalues move down the imaginary axis and are absorbed into the lower half ξ plane, or pairs of eigenvalues collide and bump each other off the imaginary axis. When the latter happens a pair of nonimaginary eigenvalues is formed, located symmetrically about the imaginary axis. For example, there could be two eigenvalues $\xi_m(C) = \omega_m(C) + i\eta_m(C)$, $m = 1, 2$, that are on the imaginary axis for $0 < C < C_0$ but that collide at $C = C_0$ and remain symmetric about the axis for $C > C_0$. Thus, the real and imaginary parts would satisfy

$$\begin{aligned} \omega_m(C) &= 0, & \eta_1(C) &\neq \eta_2(C), & 0 < C &< C_0, \\ -\omega_1(C) = \omega_2(C) &> 0, & \eta_1(C) &= \eta_2(C), & C &> C_0. \end{aligned}$$

As discussed below (5.29), the initial pulse evolves (as $\zeta \to \infty$) into an order $N = 2$ soliton for $0 < C < C_0$. For $C > C_0$, however, the terms in (5.28) corresponding to the nonimaginary eigenvalues represent traveling waves that separate in time (τ) as $\zeta \to \infty$. The chirp level $C = C_0$ therefore constitutes a soliton destabilization threshold. In other words, one can apply so much chirp that the input pulse (5.37) does not evolve into a soliton even if there is sufficient energy for eigenvalues to exist. Conditions on threshold energy for soliton formation will be considered in the next section.

Overchirp destabilization can occur even if there is a single eigenvalue. For some pulse shapes (but not all) there can be a single eigenvalue associated with (5.29) for $C = 0$, which begins to move as C increases. But then a "rogue" eigenvalue emerges from the lower half-plane and collides with the first eigenvalue to form a symmetric nonimaginary pair [KS1]. This peculiar characteristic seems to be associated with pulse shapes $q_0(\tau)$, which have steep shoulders and flat central regions. Mathematical conditions that guarantee the existence of attack eigenvalues are not known. There are no killer eigenvalues for Gaussian or hyperbolic secant pulses, but rectangular pulses of sufficient area appear always to have colliding eigenvalues. For a rectangular pulse of width 2 ($-1 \leq \tau \leq 1$) the height must be at least $\pi/4$ in order for an eigenvalue to exist. If the height H satisfies $\pi/4 < H < 1.93$, then the single eigenvalue moves down the imaginary axis and is absorbed as C increases. If $H > 1.93$, then all eigenvalues are bumped off the imaginary axis by another eigenvalue. If there is an even number of eigenvalues, they collide pairwise; if there is an odd number, then the lowest eigenvalue is chased down by an attacker while the remaining ones collide pairwise. For details see [KS1]. Collisions, and especially attacker/killer eigenvalues, are very interesting phenomena that could have physical implications.

One can ask whether increasing C indefinitely eradicates all eigenvalues eventually. This is true and is shown in [KS1].

5.3 Thresholds for Eigenvalue Formation

As mentioned, for the real symmetric rectangular pulse shape, the total area under the curve must exceed $\pi/2$ in order for (5.29) to have an eigenvalue. In general one can ask if there is a threshold level L for the L_1 norm of $q(\tau)$ in (5.29),

$$E = \int_{-\infty}^{\infty} |q(\tau)| d\tau, \tag{5.38}$$

such that if $E < L$, then there are no eigenvalues. The best possible constant L was determined recently [KS3] and found to be exactly $\pi/2$. Previously the best bound was $E < 1.32$ [NMPZ]; the bound $E < 0.904$ was given in [AbS]. A complete statement of the $\pi/2$ theorem and an outline derivation will be given in this section.

Note that the expression (5.38) is not energy since $|q(\tau)|$ is the square root of power; see section 1.1. Nevertheless, (5.38) can be taken as an indicator of total pulse energy.

The main result of [KS3] is the following.

Theorem A. *Let $q(\tau)$ be a real-valued L_1 function ($E < \infty$) and define*

$$I = \int_{-\infty}^{\infty} q(\tau) d\tau.$$

Let N be the largest nonnegative integer such that $|I| > (2N-1)/2$. Then (5.29) has at least N purely imaginary eigenvalues. In particular, if $|I| > \pi/2$, then there is at least one purely imaginary eigenvalue.

To get an idea of the elements of a proof, suppose that $q(\tau)$ has compact support $[-d, d]$. Note that $\xi = is$ is an eigenvalue with eigenfunction $\vec{v}(\tau)$ if and only if $v_1(d) = v_2(-d) = 0$ by compactness of support. Additionally, $v_1(-d) = 1$ can be taken as a normalization. Define the Prüfer substitution

$$\begin{pmatrix} v_1 \\ v_2 \end{pmatrix} = \begin{pmatrix} \rho\cos(\theta) \\ \rho\sin(\theta) \end{pmatrix} \tag{5.39}$$

and note that at an eigenvalue one has $\theta(d, s) = (2k-1)\pi/2$. A technical lemma is that $\theta(d, s) \to 0, s \to \infty$ [KS3]. If (5.39) is substituted into (5.29), one result is $\theta' = -q(\tau) - s\sin(2\theta)$. For $s = 0$ this integrates to $\theta(d, 0) = -\int_{-d}^{d} q(\tau)d\tau$ and thus $|\theta(d, 0)| > (2N-1)\pi/2$ by hypothesis. Since $\theta(d, s) \to 0$ as s increases without bound, then there are at least N crossings of the level $\theta(d, s) = (2k-1)\pi/2$; that is, there are at least N eigenvalues.

There may be more than N crossings for some potentials because $\theta(d, s)$ need not be monotonic in s [KS3]. However, for single lobe potentials the number of eigenvalues is exactly N [KS3].

One may ask if Theorem A has a converse, that is, whether $|I| \leq \pi/2$ rules out eigenvalues. This is indeed the case, as seen from the next result [KS3]. First, define

$$q_+(\tau) = \max[q(\tau), 0], \qquad q_-(\tau) = \max[-q(\tau), 0].$$

Theorem B (see [KS3]). *Let $q(\tau)$ be a real-valued L_1 function such that*

$$\int_{-\infty}^{\infty} q_+(\tau)d\tau \leq \pi/2 \quad \textit{and} \quad \int_{-\infty}^{\infty} q_-(\tau)d\tau \leq \pi/2.$$

Then there are no purely imaginary eigenvalues. In particular, there are no purely imaginary eigenvalues if

$$\int_{-\infty}^{\infty} |q(\tau)|d\tau \leq \pi/2. \tag{5.40}$$

Combining Theorem A, Theorem B, and the results from section 5.2, one now has the following definitive classification on eigenvalue thresholds for single lobe potentials.

Theorem C. *Let $q(\tau)$ be a single lobe L_1 function. If N is the largest nonnegative integer such that $I > (2N-1)/2$, then* (5.29) *has exactly N purely imaginary eigenvalues and none that are nonimaginary.*

Results on eigenvalue location for more general L_1 potentials $q(\tau)$ are available. It is shown in [KS3] that if $q(\tau)$ is of a single sign and (5.29) has eigenvalues at all, then there is a purely imaginary eigenvalue whose magnitude is strictly larger than the imaginary part of any other eigenvalue. As a consequence [KS3], if (5.29) with $q(\tau)$ replaced by $|q(\tau)|$ has no imaginary eigenvalues, then (5.29) (with $q(\tau)$) has no eigenvalues.

Finally, criterion (5.40) can be extended to complex-valued L_1 potentials. This is recorded in the following.

Theorem D (see [KS3]). *Suppose that $q(\tau)$ is complex valued, in L_1, and satisfies* (5.40). *Then* (5.29) *has no eigenvalues.*

5.4 Remarks and Summary

This chapter has attempted to profile the main points in the vast area of optical solitons. These remarkable solutions of the NLSE (5.1) are both intrinsically interesting and physically relevant to fiber communications. The link between the NLSE and the IST method is an ideal transfer point between engineering mathematics and pure analysis.

Solitons exist in contexts other than (5.1), solitary wave solutions of the coupled Manakov equations being one example [Ag1]. In the Manakov case, it is known that explicit soliton solutions exist, but the IST formalism has not been worked out in most cases [Ag1]. Closed form hyperbolic secant solutions appear to exist for coupled systems of equations with arbitrarily many components [YeB]. If the term containing $\frac{\partial(|A|^2A)}{\partial t}$ in (3.46) is added to (3.1), the resulting equation is integrable by inverse scattering [KoV].

The chapter now closes by expressing the soliton threshold results of section 5.3 in terms of physical units (rather than dimensionless ones) bringing the reader full circle, as it were, back to the beginning of the book.

The starting point is the NLSE (3.1), written here as

$$iA_z = (\beta_2/2)A_{TT} - \gamma|A|^2 A, \qquad A(0,T) = f(T), \tag{5.41}$$

where z, T, and $f(T)$ are physical distance, local pulse time, and a given initial pulse shape, respectively. Solitons can exist only in the anomalous dispersion regime $\beta_2 < 0$ (section 4.3). The form (5.1) results from the changes of variables at the end of section 3.4. In terms of the new variables the initial value $A(0,T) = f(T)$ becomes $q(\tau) = u(0,\tau) = Nf(T)/\sqrt{P_0}$, where $P_0 = |A(0,0)|^2$ is the peak power. Therefore, I in Theorem A is

$$I = \int_{-\infty}^{\infty} q(\tau)d\tau = \frac{N}{\sqrt{P_0}}\int_{-\infty}^{\infty} f(T)\frac{dT}{T_0} = \frac{N}{T_0\sqrt{P_0}}\int_{-\infty}^{\infty} f(T)dT. \tag{5.42}$$

However, $N^2 = T_0^2\gamma P_0/|\beta_2|$, and so cancellation gives

$$I = \sqrt{\frac{\gamma}{|\beta_2|}}\int_{-\infty}^{\infty} f(T)dT.$$

For single lobe potentials in (5.3), the criterion $\pi/2 < \int_{-\infty}^{\infty} q(\tau)d\tau < 3\pi/2$ for the existence of a single eigenvalue, and consequently the emergence of the fundamental soliton, is thus

$$\frac{\pi}{2} < \sqrt{\frac{\gamma}{|\beta_2|}}\int_{-\infty}^{\infty} f(T)dT < \frac{3\pi}{2},$$

a condition that depends only on the physical constants in the governing equation (5.41) and the initial pulse shape.

For example, if $q(\tau) = u(0,\tau) = N\,\mathrm{sech}(\tau)$, as below (5.1), then $\int_{-\infty}^{\infty}\mathrm{sech}(\tau)d\tau = \pi$ gives the well-known criterion $1/2 < N < 3/2$ [Ag1]. If $f(T)$ is the rectangular shape of height $\sqrt{P_0}$ between $-T_0$ and T_0, then (5.42) computes as $2N$ and so the fundamental soliton in Figure 5.1 emerges if $\pi/4 < N < 3\pi/4$. For Gaussian pulses,

$$f(T) = \sqrt{P_0}\exp\left\{-\frac{T^2}{2T_0^2}\right\},$$

the corresponding criterion is $\sqrt{\pi/8} < N < 3\sqrt{\pi/8}$. Numerically this is

$$0.6267 < N < 1.8800. \tag{5.43}$$

In physical units $\sqrt{\pi/8} < N < 3\sqrt{\pi/8}$ is

$$\sqrt{\pi/8} < \sqrt{\frac{T_0^2\gamma P_0}{|\beta_2|}} < 3\sqrt{\pi/8}. \tag{5.44}$$

Note that (5.44) induces a family of curves indexed by $N = \sqrt{\frac{T_0^2\gamma P_0}{|\beta_2|}}$. For example, one could adjust the width T_0 and compensate the center height P_0 without changing N.

Going back to the Gaussian variational approximation in Chapter 4, Subcase (iia) in section 4.3.6, the condition $(1/\sqrt{2}) < s_0^2 < \sqrt{2}$ induces the formation of optical solitons, within the Gaussian approximation. Recalling $s_0 = N$, $(1/\sqrt{2}) < s_0^2 < \sqrt{2}$ converts to $0.8409 < N < 1.1892$, as compared to the exact condition (5.43). The variationally calculated range of N is therefore correct but is overly conservative.

Bibliography

[AbS] M. J. Ablowitz and H. Segur, *Solitons and the Inverse Scattering Transform*, SIAM Studies in Applied Mathematics 4, SIAM, Philadelphia, 1981.

[Ada] M. Adams, *An Introduction to Optical Waveguides*, John Wiley & Sons, New York, 1981.

[Ag1] G. P. Agrawal, *Nonlinear Fiber Optics*, 3rd Ed., Academic Press, New York, 2001.

[Ag2] G. P. Agrawal, *Fiber-Optic Communication Systems*, John Wiley & Sons, New York, 1992.

[AkA] N. N. Akhmediev and A. Ankiewicz, *Solitons: Nonlinear Pulses and Beams*, Chapman and Hall, London, 1997.

[AKNS] M. Ablowitz, D. Kaup, A. Newell, and H. Segur, The inverse scattering transform-Fourier analysis for nonlinear problems, *Stud. Appl. Math.* 53 (1974), 249–315.

[And] D. Anderson, Variational approach to nonlinear pulse propagation in fibers, *Phys. Rev. A* 27 (1983), 3135–3145.

[BC1] R. Beals and R. R. Coifman, Scattering and inverse scattering for first order systems, *Comm. Pure Appl. Math.* 37 (1984), 39–90.

[BC2] R. Beals and R. R. Coifman, Inverse scattering and evolution equations, *Comm. Pure Appl. Math.* 38 (1985), 29–42.

[BDT] R. Beals, P. Deift, and C. Tomei, *Direct and Inverse Scattering on the Line*, Mathematical Surveys and Monographs, No. 29, American Mathematical Society, Providence, RI, 1988.

[BJL] M. Brandt-Pearce, I. Jacobs, J.-H. Lee, and J. K. Shaw, Variational analysis of optimal pulse widths in dispersive nonlinear fibers, in *Proceedings of Conference on Lasers and Electro-Optics* (CLEO 99), paper CThK60, Optical Society of America, Washington, DC, 1999, 436–437.

[BlW] K. Blow and D. Wood, Theoretical description of transient stimulated Raman scattering in optical fibers, *IEEE J. Quantum Electronics* 25 (1989), 2665–2673.

[Bou] J. Bourgain, *Global Solutions of Nonlinear Schrödinger Equations*, American Mathematical Society, Providence, RI, 1999.

[BPJ] M. Brandt-Pearce, I. Jacobs, J.-H. Lee, and J. K. Shaw, The optimal input Gaussian pulse width for transmission in dispersive nonlinear fibers, *J. Opt. Soc. Amer. B* 16 (1999), 1189–1196.

[BuC] P. N. Butcher and D. Cotter, *The Elements of Nonlinear Optics*, Cambridge University Press, Cambridge, UK, 1990.

[Den] D. Denby, The speed of light, *The New Yorker*, November 2000, 132–141.

[DeZ] P. Deift and X. Zhou, Long-time asymptotics for solutions of the NLS equation with initial data in a weighted Sobolev space, *Comm. Pure Appl. Math.* 56 (2003), 1029–1077.

[EcV] W. Eckhaus and A. Van Harten, *The Inverse Scattering Transform and the Theory of Solitons*, North-Holland, Amsterdam, 1981.

[Fra] P. L. Francois, Nonlinear propagation of ultrashort pulses in optical fibers: Total field formulation in the frequency domain, *J. Opt. Soc. Amer. B* 8 (1991), 276–293.

[Geo] T. Georges, Soliton interaction in dispersion-managed links, *J. Opt. Soc. Amer. B* 15 (1998), 1553–1560.

[Glo] D. Gloge, Weakly guiding fibers, *Applied Optics* 10 (1971), 2252–2258.

[Gof] D. Goff, *Fiber Optic Reference Guide*, 2nd Ed., Focal Press, Boston, 1999.

[GrY] V. S. Grigorian, T. Yu, E. A. Golovchenko, C. R. Menyuk, and A. N. Pilipetskii, Dispersion-managed soliton dynamics, *Opt. Lett.* 22 (1997), 1609–1611.

[Hec] J. Hecht, *City of Light, the Story of Fiber Optics*, Oxford University Press, Oxford, UK, 1999.

[IEC] The International Engineering Consortium, *The Basics of Telecommunications*, presented by the International Engineering Consortium (IEC), Professional Education International, Chicago, 2002.

[Ish] A. Ishimaru, *Electromagnetic Wave Propagation, Radiation and Scattering*, Prentice-Hall, Englewood Cliffs, NJ, 1991.

[JaS] I. Jacobs and J. K. Shaw, Optimal dispersion pre-compensation by pulse chirping, *Applied Optics* 41 (2001), 1057–1062.

[JHL] J.-H. Lee, *Analysis and Characterization of Fiber Nonlinearities with Deterministic and Stochastic Signal Sources*, Ph.D. Dissertation, Virginia Polytechnic Institute and State University, Blacksburg, VA, 1999.

[JLS] I. Jacobs, J.-H. Lee, and J. K. Shaw, Spectral asymptotics of nonlinear pulse propagation in fibers, *SPIE Proceedings* 3491 (1998) (ICAPT 98, July, Ottawa, Ontario, Canada), 538–543.

[JSW1] I. Jacobs, J. K. Shaw, and N. Wongsangpaiboon, Variational approach to optimal dispersion compensation, *SPIE Proceedings* 4087 (2000) (ICAPT 2000, July, Quebec City), 420–428.

[JSW2] I. Jacobs, J. K. Shaw, and V. Wongpaibool, Effect of fiber loss on optimal signal design for dispersive nonlinear fibers, in *Proceedings of Conference on Lasers and Electro-Optics* (CLEO 2001), Optical Society of America, Washington, DC, 2001, 473–474.

[KaM] D. J. Kaup and B. A. Malomed, The variational principle for nonlinear waves in dissipative systems, *Physica D* 87 (1995), 155–159.

[KHEG] N. Kutz, P. Holmes, S. Evangelides, and J. Gordon, Hamiltonian dynamics of dispersion-managed breathers, *J. Opt. Soc. Amer. B* 15 (1998), 87–96.

[KiSt] Y. Kivshar and G. Stegeman, Spatial optical solitons, *Optics and Photonics News* 13 (2002), 59–63.

[Kot] V. P. Kotlyarov, Asymptotic analysis of the Marchenko integral equation and soliton asymptotics of a solution of the nonlinear Schrödinger equation, *Physica D* 87 (1995), 176–185.

[KoV] M. Kovalyov and A. Vartanian, On long-distance intensity asymptotics of solutions to the Cauchy problem for the modified nonlinear Schrödinger equations for vanishing initial data, in *Recent Developments in Integrable Systems and Riemann–Hilbert Problems*, Contemporary Mathematics 326, K. McLaughlin and X. Zhou, Eds., American Mathematical Society, Providence, RI, 2003, 49–57.

[KS1] M. Klaus and J. K. Shaw, Influence of pulse shape and frequency chirp on stability of optical solitons, *Optics Communications* 197 (2001) 491–500.

[KS2] M. Klaus and J. K. Shaw, Purely imaginary eigenvalues of Zakharov-Shabat systems, *Phys. Rev. E* 65 (2002), paper 036607, 1–8.

[KS3] M. Klaus and J. K. Shaw, On the eigenvalues of Zakharov–Shabat systems, *SIAM J. Math. Anal.* 34 (2003), 759–773.

[KuE] N. Kutz and S. Evangelides, Dispersion-managed breathers with average normal dispersion, *Opt. Lett.* 23 (1998), 685–687.

[Lam] G. L. Lamb, *Elements of Soliton Theory*, John Wiley & Sons, New York, 1980.

[LiC] W. Liu and S. Chui, Integrable models in Bose-Einstein condensates, in *Recent Developments in Integrable Systems and Riemann–Hilbert Problems*, Contemporary Mathematics 326, K. McLaughlin and X. Zhou, Eds., American Mathematical Society, Providence, RI, 2003, 59–90.

[Log] D. Logan, *Applied Mathematics*, 2nd Ed., John Wiley & Sons, New York, 1997.

[Mal] B. A. Malomed, Variational methods in nonlinear fiber optics and related fields, *Progress in Optics* 43 (2002), 71–193.

[Mar] D. Marcuse, *Light Transmission Optics*, Van Nostrand Reinhold, New York, 1982.

[MBES] A. Mainistov, A. Basharov, S. Elyutin, and Y. Sklyarov, Present state of self-induced transparency theory, *Physics Reports* 191 (1990), 1–108.

[Men] C. Menyuk, Application of multiple length scales to the study of optical fiber transmission, *J. Engineering Mathematics* 36 (1999), 113–136.

[Mil] D. L. Mills, *Nonlinear Optics*, Springer-Verlag, Berlin, 1991.

[NeM] A. Newell and J. Maloney, *Nonlinear Optics*, Addison-Wesley, Redwood City, CA, 1992.

[NMPZ] S. Novikov, S. V. Manakov, L. P. Pitaevskii, and V. E. Zakharov, *Theory of Solitons: The Inverse Scattering Method*, Consultants Bureau, New York, 1984.

[Sav] N. Savage, The dawn of fiber optics, *OPN Trends* 1 (2001), 20–22.

[Sch] P. C. Schuur, *Asymptotic Analysis of Soliton Problems*, Lecture Notes in Mathematics 1232, Springer-Verlag, Berlin, 1986.

[SHW] J. K. Shaw, W. Henry, and R. Winfrey, Weakly guiding analysis of elliptical core step index optical waveguides based on the characteristic numbers of Mathieu's equation, *J. Lightwave Technology* 13 (1995), 2359–2371.

[XMa] X. Ma, *Analysis and Simulation of Long Haul In-Line Fiber Amplifier Transmission Systems*, Ph.D. Dissertation, Virginia Polytechnic Institute and State University, Blacksburg, VA, 1994.

[YeB] C. Yeh and L. A. Bergman, Existence of optical solitons on wavelength division mutliplexed beams in a nonlinear fiber, *Phys. Rev. E* 60 (1999), 2306–2308.

[ZaS] V. E. Zakharov and A. B. Shabat, Exact theory of two-dimensional self-focusing and one-dimensional self-modulation of waves in nonlinear media, *Sov. Phys. JETP* 34 (1972), 62–69.

[Zh1] X. Zhou, Zakharov-Shabat inverse scattering, in *Scattering: Scattering and Inverse Scattering in Pure and Applied Science*, R. Pike and P. Sabatier, Eds., Vol. 2, Chapter 6.2.2, Academic Press, San Diego, 2002, 1707–1716.

[Zh2] X. Zhou, Direct and inverse scattering transforms with arbitrary spectral singularities, *Comm. Pure Appl. Math.* 42 (1989), 895–938.

Index

Printed in the United States
17737LVS00002B/139-510